# APPARENTLY FLAT — BUT NOT

# APPARENTLY FLAT — BUT NOT

How Appearances Mislead

and

Measurements Reveal the Shape of the Earth

Will York

Disclaimer:

This book presents arguments for evaluation and encourages readers to independently verify the claims and measurements discussed.

This book is not intended to provide professional, scientific, medical, or legal advice. The views expressed are those of the author and are presented for the purpose of critical examination and discussion.

First Edition

Published in the United States of America by Ideaphoria, LLC

Cover photograph by hangphe*, Adelaide via pexels.com

To those willing to question
and follow evidence
wherever it leads
and who challenge
prevailing narratives
in the search for truth.

---

Dedicated with sincere appreciation to

Joe Rambo of *RamboAndFriends*

Dave of *News Treason (Channel 17)*

Paul Fleuret & Stormy Patriot Joe of *Absolute Storm*

Patrick Gunnels of *pgunnels* and *tulipblooms.shop*

Mickey and Matt of *Above The Rabbit Hole*

Mike King of *RealNewsandHistory.com*

BB Maguire of *Doses of Deception*

*Mark (8traQ) of Parallel 8 Media*

JFanon, Sapioplex, the Zoids

*Patriot Underground*

*MatrixQdeBreaker*

*qagg.news*

The Boss

**Q**

# Table of Contents

# INTRODUCTION

## Apparently Flat — But Not

I consider myself to be what many would call a conspiracy theorist.

I am convinced that there are people with power, money, and influence who have secretly cooperated—conspired—to protect and expand their power by deceiving others about important aspects of reality. I am careful to say I am *convinced*, not *certain*. I do not claim infallible knowledge. I have heard the mainstream narratives, examined the reasons offered for them, listened carefully to alternative explanations, and concluded that many official accounts are false or seriously misleading.

Since I do not want to believe falsehoods, I remain deliberately open to the possibility that some of my beliefs are wrong. I try to welcome questions and challenges because I believe the best way to expose false beliefs—especially my own—is to allow others to test my reasons for believing what I do.

Over the years, this posture has led me to become open to—and to eventually accept—a number of conspiracy explanations for major events and systems in the modern world. Broadly speaking, these fall into several overlapping categories.

First, political and historical conspiracies. I am convinced that many wars were not accidents or tragic inevitabilities, but were deliberately instigated by powerful banking, financial, and media interests who stood to benefit from conflict. I am convinced that the events of September 11, 2001, were planned long in advance and intentionally blamed on Islamic terrorists to justify decades of war and the expansion of state power.

Second, domestic control and disarmament narratives. I am convinced that many high-profile mass shooting events in the United States—such as Columbine, Sandy Hook, Parkland, and the Las Vegas country music festival—were false-flag operations designed to manipulate public opinion against gun ownership and to justify disarming the population. I believe these efforts were part of a broader attempt to weaken the ability of ordinary people to resist tyranny.

Third, financial and institutional conspiracies. I am convinced that systems like the Federal Reserve and the IRS were deliberately designed to enslave the population through debt, to extract wealth upward, and to impoverish nations over time. I do not believe the enormous national debt we face is accidental.

Fourth, medical, pharmaceutical, and food-industry conspiracies. I am convinced that large pharmaceutical companies have conspired to deceive the public about the causes of disease and the nature of health. I do not believe viruses are the primary cause of disease. I'm convinced viruses—as they are commonly described—do not exist. I believe toxicity, nutrient deficiencies, trauma, parasites, and the effects of highly processed foods play a central role in disease. I am convinced that major food corporations have paid for fraudulent research, lobbied governments to restrict access to genuinely healthy food, and helped create a cycle in which poor nutrition leads to chronic illness that is then "managed" with profitable medications.

Fifth, educational and ideological conspiracies. I am convinced that modern education systems have deliberately indoctrinated children with false views of reality—particularly materialistic evolution—while withholding instruction in logic, practical skills, financial literacy, self-improvement, and how to rigorously test ideas using language and reason.

Underlying all of these beliefs is a more unsettling conviction: that some people who possess enormous power genuinely enjoy doing evil and are willing to cause widespread suffering, chaos, and destruction in order to maintain control.

I include all of the above not as conclusions I expect you to accept, but only to make clear the perspective I bring into this discussion. I am always open to changing my beliefs about them.

Because of these convictions, I spend much of my time listening to alternative media and independent voices who challenge official narratives. Since the COVID "plandemic"—which I believe was itself a coordinated operation to harm populations worldwide—I have followed and listened to many conspiracy-focused channels and commentators, including Derek Johnson of thedocuments.info, RamboAndFriends, Patriot Underground, News Treason (Channel 17), Above The Rabbit Hole, pgunnels, Parallel 8 Media, Mike King of RealNewsandHistory.com, and Doses of Deception. I have read many of Mike King's history books and regularly listen to current-event discussions hosted on platforms such as Rumble, BitChute, Truth Social, YouTube, X, and Pilled.

(I encourage you to give the above channels on Rumble a listen and see if you enjoy their pursuit of truth. Also, check out the two mentioned websites and be prepared to have your beliefs tested.)

As a result of following these voices, I am convinced that a global "white-hat" military operation is underway to dismantle what is often called the Deep State—an entrenched network of elites who have exercised covert control over governments, institutions, and populations for centuries. At the time of writing, I describe myself as a "full control" believer: I believe that militaries are controlling global information flow, that much of what we see in the media is scripted or staged, and that the public is being slowly prepared, through carefully timed disclosures, for a massive release of truth.

I include all of the above for one reason: I am not someone who reflexively trusts authority, institutions, or official explanations. Quite the opposite. My default posture is suspicion. My bias is toward conspiracy explanations, not against them.

That fact makes the subject of this book especially important.

Among conspiracy-minded communities, few ideas are as central—or as identity-forming—as the belief that the Earth is flat. Flat Earth is often presented not merely as a scientific claim, but as one of the grandest deceptions conspirators have orchestrated: a lie that makes many other deceptions possible. If the Earth is flat, then governments, scientists, educators, and media organizations across the world must be colluding to hide that fact.

If that claim were true, someone like me should be among the easiest people to convince.

For my entire life, however, I have believed the Earth is a sphere. That belief has been one of my strongest and least questioned assumptions—until I began hearing Dave of News Treason and Patrick Gunnels claim that "simple math" disproves the spherical Earth model altogether. When I first heard that claim, it genuinely shook me. I remember thinking: "Do these guys know something I don't? My beliefs could be wrong. If the math really does rule out a sphere, I will have to follow that evidence wherever it leads." And I hope you will think that way as you read this book.

When I began exploring non-spherical content, I quickly recognized serious errors in many of the arguments. Since I felt certain the Earth is a sphere, I thought about ignoring the debate entirely. But I continued to hear the claim repeated—the math doesn't work—and I realized I needed to test that claim carefully.

On January 12, 2026, the channel *Above The Rabbit Hole* aired a discussion titled "Pat Talks Flat Earth with Mickey." ("Pat" referring to Patrick Gunnels.) Listening to Pat carefully, I immediately recognized patterns of reasoning that I believed were flawed, along with misunderstandings of physics, optics, perspective, air pressure, and misuse of an otherwise proper argument. In the 1990s, I taught high school biology and chemistry, so I am a science teacher by background and enjoy understanding and explaining how the physical world works. I saw problems, but I wanted to test my beliefs. Since I respect Pat's intelligence and sincerity, I reached out to him—not to debate publicly, but to discuss the evidence. My goal was not to win an argument or persuade him to agree. It was to examine the evidence with him and see what is true. He replied that he would be willing to have a discussion whenever I was ready to do so.

Preparing for that discussion, which ultimately did not take place, led me to an unexpected tool: ChatGPT. I used it to ask basic questions about Earth's size, curvature, and the mathematics involved. It walked through the "simple math" that I'd heard Pat and Dave claim made it impossible for the Earth to be a sphere, and it then offered to explain the equation surveyors use to correct for refraction. I knew about refraction, but I was not aware of an equation to correct for it, so I typed, "yes please." After reading the answer and asking additional questions, I realized two things: first, that the mathematics for determining visibility are not as "simple" as Pat and Dave have suggested; and second, that they may have reached their flat-Earth conclusions without taking into account all the variables involved in what makes things apparent at great distances over the Earth's surface.

Those realizations made me start thinking about how appearances can be deceptive, and how measurements can help us confirm or correct our perceptions.

At first, I planned to write Pat an email explaining why appearances must be verified with measurements. But it quickly became clear that an email would not be enough, so I decided to write him this book.

I then went to the largest flat-Earth website I could find—flatearthdave.com—and listened to multiple podcasts and videos. What I quickly realized was not only that some flat-Earth arguments are incomplete, but that several commonly repeated claims are either logically unsound, factually false, or both.

I asked ChatGPT more about surveying and then about other technologies that use measurements and suggest the Earth is a sphere. Its answers led me to notice something I had previously taken for granted: many of the technologies we use every day do not function without relying on mathematical systems and measurements that both assume and confirm that the Earth is a sphere. Navigation systems, surveying networks, radio coverage maps, and global positioning do not succeed because people "believe the right thing," and they are not maintained by people conspiring to defend a theory. They are built by engineers who are designing reliable systems that millions of customers depend on every day. The shape of the Earth is embedded directly in the calculations behind technologies that many of us take for granted. Each of these claims is worth testing by further examination.

The point is this: these technologies succeed because their equations continue to match what instruments measure in the real world—day after day—reflecting the Earth's spherical geometry. If the spherical geometry they use were incorrect, these systems would drift, accumulate error, and eventually fail in practical, observable ways that would significantly disrupt our modern way of living.

Scientific models are not judged by a single appearance, argument, or prediction in isolation. They are judged by their ability to consistently explain measurements across many independent systems. This ability is often called "explanatory power." When tested against measurements from surveying, navigation, radio communication, and Global Positioning Systems, the model that makes sense is the spherical model—not non-spherical models, which tend to rely primarily on how the Earth appears from human perspective rather than on tool-obtained measurements.

In our ordinary human experience, the Earth often appears flat to us, and I readily acknowledge that fact. But the main question I ask herein is whether those appearances alone can reliably tell us the true shape of physical reality. In this book, I argue that they cannot.

Through careful definitions, tool-obtained measurements, repeated observations, and cross-checked mathematical reasoning, I will explain why the Earth can appear flat and yet still be a sphere—and why many non-spherical arguments fail, not because their proponents are foolish or dishonest, but because appearances are perspective-based and therefore more deceptive than we intuitively realize.

If you are someone who distrusts authority, questions official narratives, and believes conspiracies are often real, then we have that in common, and you are exactly the kind of reader I had in mind when writing this book.

I am not an authority on the systems discussed in this book. Professionals working with these systems explain these measurements in detail online, and I encourage you to verify these ideas for yourself rather than simply accepting my descriptions.

(*Note on usage: In this book, "Earth" is used as a proper name for the planet, consistent with its treatment alongside other celestial bodies. The form "the Earth" is often used where it improves readability.*)

Turn the page, and let's examine the evidence carefully—together.

# Chapter 1

## Apparently Flat

There is no doubt about it: from our perspective, Earth appears flat.

The debate over Earth's shape begins with this simple observation. Nothing in our daily experience requires us to imagine the surface curving down beneath our feet.

Everywhere we live, work, and travel, the ground beneath us feels as though it rests on a level and stationary foundation. We walk across fields, pour foundations, and build roads, bridges, and cities as if the land extends outward in a plane, not over a curve.

Bodies of water appear flat. Lakes settle into smooth, level surfaces. Oceans spread out evenly, and instead of seeing them slope away from us toward a vanishing point where things simply become too small to see, the horizon appears to rise to our eye level. Canals run for miles without visible rise or fall. The Nile River itself drops only a few feet in elevation relative to sea level over thousands of miles, which seems to make it essentially level with the sea. The entire surface of Lake Michigan has been measured at approximately 574 feet above sea level in all directions, which can seem to imply that its surface is not curving. These are not abstract theories; they are features of the world around us that we interact with every day.

When we look out across great distances, we see rolling hills and mountains that appear smaller the farther away they are, but they do not seem to disappear over a curve as though the Earth were a giant sphere. Railways do not arc. Causeways do not crest and dip. Horizons remain at about eye level and appear as straight, unbending lines rather than curves. In many places, distant terrain, buildings, and coastlines remain visible far beyond what simple calculations tell us should be possible. If the surface curves downward between observer and object, those features should be hidden. Objects that should be completely hidden remain visible and can often be seen in full with telescopes or optical zoom lenses.

These geometric calculations are straightforward and accessible. Geometry is the branch of mathematics concerned with measurable spatial relationships such as distance, angle, position, direction, and curvature. If the Earth is a sphere, then the geometry of its surface should have a downward curvature of approximately eight inches per mile squared—roughly sixty-six feet every ten miles. When observers report seeing Chicago across Lake Michigan from roughly sixty miles away, or the island of Oahu from Kauai at a distance of about one hundred eight miles, the conclusion can feel unavoidable. When that predicted curvature does not appear in real-world observations across land and water, it can feel as though the spherical model must be wrong. If the Earth truly curved downward at eight inches per mile squared, it seems we simply should not be able to see what we can see at such great distances. It is natural, then, to ask: what other proof could be needed?

Our everyday experience of non-movement seems to confirm what we see. We are told by scientists that Earth is spinning at thousands of miles per hour and flying through space at even greater speeds, yet we feel no rotation, no rushing of air, no sideways pull, and no direct evidence that the ground beneath us is moving. Nothing in our immediate observation suggests that we are standing on a spinning sphere hurtling through space at tremendous velocity.

Water appears to confirm this same picture. Wherever we encounter water on Earth, it appears flat and level. At an intuitive level, when sphere-shaped objects spin, water does not cling to them—it flies off. Anyone who has spun a wet ball or watched liquid leave a rotating surface finds this analogy compelling. Yet Earth's oceans and lakes remain in place. Water finds its level and stays there. It does not seem to behave as though it is attached to a rapidly spinning sphere.

The sky appears to tell a similar story. The Sun rises, moves across the sky, and sets. The stars trace predictable paths above us. Polaris remains fixed. From where we stand, the heavens appear to move while Earth does not. Time-lapse photography seems to confirm that the lights in the sky move in circles above an unmoving ground. Night after night, this pattern repeats itself with remarkable consistency.

Trusting what we experience with our own senses—and what others show us through video recordings—is how we ordinarily navigate the world we inhabit. When we watch clear videos recorded by ordinary people performing observations that many of us would gladly perform ourselves if we lived in the right location, shouldn't we take that evidence seriously?

Amateur videographers do not possess the resources of government agencies to fabricate or stitch together artificial images of a spherical Earth, which those agencies themselves acknowledge they create.

Taken together, our senses—sight, touch, balance—and simple calculations can powerfully suggest that the Earth is flat and unmoving. When all of these seem to point in the same direction, believing the Earth is flat and not moving can feel like the only reasonable conclusion.

Although I have just described the Earth as apparently flat and unmoving, those are two separate topics. One concerns the shape of the Earth. The other concerns whether the Earth moves through space or remains stationary. This book addresses only the shape of Earth. Earth's motion deserves its own careful investigation.

In most situations, I think it is natural and safer to trust what we see and feel more than what others tell us about reality. But when it comes to something as large as the Earth, should we always trust appearances? Can our finite senses reveal the shape of something so large that we cannot see it in its entirety? Is it possible that our senses could mislead us about aspects of physical reality that extend far beyond everyday scale? In short, are appearances always trustworthy?

History gives us good reason to pause before answering those questions too quickly. As Mark Twain famously warned, "It ain't what you don't know that gets you into trouble. It's what you know for sure that just ain't so." Again and again, human confidence in what feels obvious has been shown to be misplaced, especially when distance, size, or motion exceed the limits of direct experience.

This is not because our senses are broken or generally unreliable. In fact, they are very effective at helping us move through and interact with our immediate environment. But they were not designed to measure planetary dimensions, enormous distances, or movements that unfold across scales far beyond everyday experience.

The familiar saying "Seeing is believing" might be more accurately phrased as this: seeing feels like believing—until we start measuring.

For this reason, when questions about the physical world extend beyond everyday scales, tool-obtained measurements must take precedence over appearances, not because appearances are meaningless, but because they do not directly measure at the scale needed to test such large physical claims.

Rather than relying on sensation alone, humans have learned to investigate the physical world using tools that measure time, distance, and angles—to test appearances—and to have those measurements repeated in different places so the results can be compared, checked, and tested for consistency using the mathematics of geometry.

Human beings move through the world primarily using visual imaginations that the mind automatically constructs from sensory input. These imaginations allow us to navigate roads, recognize objects, and judge distance well enough for everyday life.

But the systems used to measure the physical world do not rely on those imaginations. They rely on geometry: the branch of mathematics concerned with measurable relationships involving distance, angle, position, direction, time, and curvature. Surveying instruments, navigation systems, satellites, and timing networks function reliably only because their measurements and calculations consistently align with the actual geometric shape of Earth.

Modern transportation, communication, and navigation technologies operate successfully only when their measurements and calculations align with the actual geometry—the actual measurable shape—of Earth. (Here and elsewhere I sometimes use "geometry" to refer to the measurable shape of something.)

When appearances are tested in this way, we sometimes discover that perspective-based imaginations created by our senses can be very clear and convincing while failing to represent reality as it really is.

If our senses can present us with mental images that feel convincing yet fail to accurately represent reality itself, then we should immediately ask ourselves how best to investigate claims about the physical world that extend far beyond the scale of everyday experience.

When the subject is something as immense as the shape of the Earth, appearances alone are not enough. What feels obvious from our limited perspective may actually conflict with what careful, repeatable measurements reveal.

Unlike our visual imaginations, measuring instruments record three fundamental quantities—time, distance, and angle—and the mathematics applied to those quantities produces the same results no matter who performs the measurements.

For this reason, the method used throughout the rest of this book will be straightforward and consistent. Rather than treating our imaginations as decisive, I will examine how appearances can be tested, checked, and either confirmed or corrected by objective, tool-obtained measurements—measurements that can be repeated by many people, compared across independent technological systems, and evaluated using mathematics and clear reasoning.

The goal is not to ignore what we see, but to place it in its proper role: something to be investigated, not something to be assumed true beyond question.

In later chapters I will encourage you to hear these ideas explained by others rather than relying only on my explanations. I encourage you to cross-check what I describe with explanations from professionals working in each field.

Before turning to measurements, we must first understand the primary sense that produces most of the appearances we rely on: vision.

# Chapter 2

## Vision, Perspective, and Illusion

The primary physical sense we rely on to tell us about the physical world is vision. Much of what we think we know about reality—distance, size, shape, motion, and position—comes to us through what we see. But can we trust our vision to give us accurate information about physical reality, especially at great distances or large scales?

Can we trust our vision to help us judge claims about the shape of the Earth? Can we trust videos provided by our peers that show us what we are told should not be visible if the Earth is a sphere? Can we trust videos streamed from the International Space Station or Artemis II? A lot of people doubt the Moon landing photos and videos are real—so can we trust any videos that NASA or mainstream media present as evidence?

As a conspiracy theorist, I am not inclined to trust anyone's videos to tell me the truth about the shape of the Earth. Videos can be manipulated. Artificial intelligence can now generate images and videos that appear convincingly real. In our current technological environment, it is increasingly difficult to verify whether a video represents an actual event or a digitally constructed image. For that reason, I do not believe it is wise to treat video evidence—whether amateur or institutional—as decisive proof of reality without independent verification.

But the problem goes deeper than AI or video manipulation. Even if a video is genuine, it still relies on the same sense that our own experience relies on: vision. And vision itself has limits.

I have a very personal reason for not trusting my eyes to give me accurate information about distance.

## A Personal Lesson About Vision and Distance

In October 2001, I was riding my mountain bike in the mountains above Santa Monica and Malibu, California. I was following my friend, who was riding about 50 yards in front of me. After descending a steep, rut-filled hill, I looked up after focusing on the ground and did not see him.

I assumed he had ridden around the next hill, which I could see was to the right of the fire road I believed he had taken. The thought went through my mind that he was going down the mountain way too fast.

I looked ahead and saw what appeared to be a small drop-off onto a gravel fire road. Since I thought my friend had ridden over it so quickly that he was already out of sight around the next hill, I assumed I could ride over the drop-off quickly too. I actually thought I needed to catch up with him and tell him to slow down so that neither of us would get hurt riding down the mountain.

At the edge of what still looked to me like a small drop-off, I bunny-hopped my bike—and discovered that my eyes had tricked me.

I was suddenly in the air, and my front wheel was about twelve feet from the far side of a small ditch at the bottom of what turned out to be about an eight-foot drop. I instantly knew that if my front wheel hit the far side of that ditch first, my face would hit second, and I would either be killed or paralyzed from the neck down.

I yelled, “Oh shit!” and tucked my head to somersault forward in an attempt to avoid landing on my face.

I landed flat on my back.

My friend later told me I was knocked unconscious for about three seconds—which was not enough time for him to reach me before I became conscious again.

When I regained consciousness, I felt tremendous pain in my back. I screamed and kicked my legs to roll onto my right side—away from the pain. The moment I rolled onto my side, my legs went numb, and I was unable to move them.

I realized I had just paralyzed myself by rolling off my back.

So, I have been paralyzed and have needed a wheelchair ever since.

I trusted my eyes and was tricked into believing the drop-off was small.

And I have suffered the consequences of that misinterpretation for more than twenty-four years.

Ouch!!

That experience does not prove anything about the shape of the Earth—but it permanently changed how I think about trusting my vision to tell me what is true.

## How Vision Works

Most of us take vision for granted without thinking about how it works. But understanding vision is essential if we want to understand its limits.

Visible light reflects off objects in the physical world and travels through the air to our eyes. Each eye functions like a simple convex lens, bending and focusing incoming light so that an image is projected onto the retina at the back of the eye. In this process, the image is flipped upside down and reversed left to right.

Cells in the retina convert incoming light into electrical signals. These signals travel along the optic nerves to the occipital lobe at the back of the brain, where they become patterns of neural activity that the mind interprets.

What happens next is what I think of as a kind of translation: the mind interprets these electrical patterns and constructs a stable, upright imagination of the outside world.

We do not see the world directly. Instead, we see mind-created images—interpretations of patterns of light detected by our eyes.

There are no pictures of trees, mountains, or horizons inside the eye or nervous system. There are only patterns of neural activity that the mind interprets. The images we experience are part of our imagination—our mind's interpretation of those signals.

In short: what we see is not reality itself—it is the mind's interpretation of incoming sensory information. Our minds construct visual imaginations designed to help us navigate our environment.

## Perspective-Based Imagination

Our visual imagination answers practical questions:

Where am I?
What is close?
What is far?
What can I interact with?

To answer these efficiently, the mind represents nearby objects as larger and more detailed, and distant objects as smaller and less detailed.

A tree that is dozens of feet tall may occupy only a tiny portion of our visual field when it is far away. As we move closer, the tree appears to grow—not because the tree itself changes, but because our position relative to it changes.

As distance increases, perspective compresses vertical differences, making surfaces appear less three-dimensional. Objects that are actually far apart can appear closer together. Parallel lines can appear to get closer together. Large features can disappear from view even though they still exist.

These effects are not defects—they are features that allow us to move through the world without constant calculation.

Obviously, our visual imaginations are extraordinarily useful. They allow us to walk, reach, throw, drive, recognize landmarks, and orient ourselves in space.

But usefulness should not be confused with accuracy.

Perspective gives us visible relationships, not measurements. It does not tell us true size, true distance, or true shape.

For this reason, questions about large structures in the physical world cannot be settled by sight alone.

## Illusions

I define an illusion as a false imagination about the outside world that the mind creates as a result of sensing real external conditions.

Illusions are not hallucinations. They arise from real sensory input, but the imagination produced by the mind does not accurately represent the underlying reality.

Several types of illusions are especially relevant to discussions about vision.

## Cognitive (Perceptual) Illusions

These occur when the mind misinterprets spatial relationships even though the light entering the eye is normal and undistorted.

Examples include:

- railroad tracks appearing to be spaced closer together at a distance
- distant mountains appearing smaller or closer together than they are
- objects appearing to rise or sink relative to the horizon
- Sun rays through clouds that appear to fan outward as they descend

Nothing is wrong with the light reaching the eye. The illusion arises from perspective compression and the mind's interpretation of spatial relationships.

## Geometric Illusions

These occur when the arrangement of lines, angles, and shapes causes the mind to misjudge geometry.

Examples include:

- straight lines appearing bent or curved
- parallel lines appearing to diverge or converge
- angles appearing larger or smaller than they are

The geometry in the world is correct. The illusion arises from how the mind interprets geometric shapes.

## Contextual Illusions

These occur when surrounding cues influence how the mind interprets spatial relationships.

This is the illusion I trusted when I misread the distance of the drop-off that I bunny-hopped before my accident. From above the drop, I did not have enough visual information to judge its true depth. There were no trees, poles, or vertical objects beyond the drop-off that could help me estimate the distance. There were only grass and a drop onto a gravel area.

In my mind's imagination, the drop-off appeared small.

Examples of contextual illusions include:

- flat surfaces appearing sloped
- objects appearing higher or lower because of background features
- objects appearing farther away because of surrounding landmarks
- colors appearing brighter or duller depending on nearby colors

The mind uses context to infer meaning—and sometimes infers incorrectly.

## Optical Illusions

Optical illusions occur when light itself is bent, scattered, reflected, or refracted before it reaches the retina.

In these cases, the mind correctly interprets the input—but the input itself is distorted. The illusion does not arise primarily from perspective or interpretation but from changes in how light travels through substances, such as water or glass, or through layers of the atmosphere before reaching our retina.

Examples include:

- mirages on hot roads caused by refracting layers of air
- a straight stick appearing bent in water due to refraction

• underwater objects appearing closer or larger than they are
• the Sun being seen for about two minutes before and after it should be visible

Note that eyeglasses work by intentionally bending incoming light to compensate for how light will be distorted by misshaped eyes, preventing false imaginations.

Many non-spherical visibility claims often mix perceptual illusion and optical distortion as if neither needs to be measured.

As I will explain, many claims presented as evidence against a spherical Earth are not new discoveries, but familiar effects that have already been carefully studied and explained.

## Why Vision Alone Is Not Enough

Given what we now understand about vision, I think we should ask ourselves: is it wise to treat visual appearances as decisive proof of physical reality, especially at great distances or large scales?

Visual appearances can be informative, but they are not measurements because vision gives us perspective-based imaginations instead of numerical quantities. What we see can suggest possibilities, but it cannot directly determine true distance, true size, or true shape.

Our vision helps us orient ourselves in the world, but by itself it cannot establish conclusions about the shape or size of the world.

For that reason, relying on vision alone to determine Earth's shape is risky. Perspective and optical effects can produce compelling appearances that feel obvious and conclusive even when they are incomplete or misleading.

To move beyond appearances, we must turn to something else: careful measurements.

Humans have developed instruments specifically designed to capture numerical measurements of reality. These measurements can be repeated, compared, and checked by many independent observers.

In later chapters we will examine how those tool-obtained measurements work and how they can be cross-checked across multiple independent systems.

But before we examine those measurements, we must first understand how claims about reality are constructed and evaluated.

If vision can mislead us this easily, then we must be careful not only when interpreting what we see, but also in how we reason with what we see.

That brings us to the subject of arguments.

# Chapter 3

## Arguments: How We Test Claims Beyond Appearances

In the previous chapter, we examined vision—how our minds construct perspective-based imaginations that are extraordinarily useful for everyday life, yet limited when we try to judge large-scale reality. Vision can mislead us about distance, size, and shape, especially when we are looking across great distances through a complex atmosphere.

But even if we fully understand the limits and misleading tendencies of vision, we still need to ask: how do we evaluate claims about reality once they are put into words?

People can look at the same horizon, the same photograph, or the same long-distance observation and draw opposite conclusions. So, if we want to determine what's true, we should focus not only on what we see, but on how we reason from what we see.

This chapter introduces the basic structure of arguments so that, in the chapters that follow, we can test claims about Earth's shape with clarity rather than confusion. Although these ideas may seem abstract at first, they become practical the moment we apply them to familiar claims about Earth's shape, so please be patient with a little repetition as I try to explain them clearly.

## What an Argument Is (and Is Not)

When philosophers use the word *argument*, they do not mean a verbal quarrel. They mean a structured set of statements in which some statements (premises) are offered as reasons for accepting another statement (conclusion).

The two basic parts of an argument mentioned above are:

- **Conclusion** — what someone is trying to establish
- **Premises** — the reasons offered in support of the conclusion

An argument can feel persuasive while still being flawed. It can be flawed because its logic is faulty, because its premises are false or incomplete, or because it quietly assumes something it never proves.

To avoid being misled, we need a few helpful terms.

## Key Parts of Real-World Arguments

### Claim

A claim is a statement that asserts something is true about reality.

For example:

"The Earth is flat."
"The Earth is a sphere."

Claims are not self-validating. All claims demand justification.

A claim is what someone is saying is the case.
It may be true or false, but by itself it is neither proven nor disproven.

We should keep that fact in mind because many discussions stall at the level of repeating claims more forcefully rather than examining whether those claims have been adequately supported.

## Support

Support is the evidence or reasons offered to justify a claim.

Support may include observations, appearances, data, calculations, measurements, or experiences. Not all "support" actually supports the claim it is meant to justify.

For example, optical illusions are not good support for a geometric conclusion. Appearances can be very convincing because we see them with our own eyes, but when they are illusions, they can lead us to wrong conclusions about the shape of the Earth.

We must keep that in mind when discussing the shape of the Earth, because many arguments rely heavily on appearances. Videos of distant cities, photographs of horizons, or personal observations across land and water may all be offered as support. But support must be examined carefully. Not all support actually supports. The fact that something is vivid, memorable, or emotionally compelling does not mean it is good evidence for the conclusion being drawn from it.

## Warrant

A warrant is the underlying assumption that connects the support to the claim.

A warrant explains why the support is supposed to prove the claim. Warrants are often unstated.

Examples include assumptions such as:

"If we can see something, then nothing is blocking it."

or

"If simple mathematics predicts something, then reality must follow that prediction exactly."

If a warrant is false or incomplete, the argument fails even if the support itself is real.

Many persuasive non-spherical arguments sound strong precisely because their assumptions are hidden. The argument may seem obvious only because the warrant has never been examined. Once the hidden assumption is brought into the open, the force of the argument often weakens immediately.

That is one reason proper argument analysis requires us to keep searching carefully for the complete truth. The most important part of many arguments is not what is said explicitly but what is assumed. Many times, in debates about the shape of the Earth, the writer or speaker assumes an idea that is foundational to their argument without expressing it, and it is our job to look for and test those hidden assumptions for accuracy.

## Backing

Backing is additional justification for the warrant.

Backing is what you appeal to when someone challenges the warrant. If the warrant depends on how light travels through air, then the backing must involve optics and refraction.

Backing is important because it separates a serious argument from mere assumption.

Without backing, a warrant may be nothing more than a guess about how the world works.

This is especially important in disputes about vision and long-distance visibility. If someone assumes that seeing a distant object proves that no curvature could be involved, that assumption needs backing. It is not enough merely to repeat the claim with confidence. One must justify why visibility should be treated as a direct test of Earth's geometry rather than as something influenced by multiple physical factors.

## Qualifier

A qualifier limits the strength or scope of a claim.

Reality is complex, so careful arguers often avoid absolutes by using phrases such as *usually*, *under typical conditions*, or *within a certain range*.

Arguments that allow no exceptions are often overstated. Good reasoning tries to account for all the relevant conditions that may support or weaken a claim.

For example, someone defending a spherical Earth might say, "Usually ships on the ocean disappear from the bottom up as they move farther away, but under some atmospheric conditions they may appear distorted, elevated, or slightly lifted before vanishing." That kind of statement is not a weakness. It is a sign that the person is trying to describe reality carefully rather than force reality to obey an overly simple rule.

## Rebuttal

A rebuttal is a response that addresses potential objections or alternative explanations.

A rebuttal strengthens an argument because it shows that alternatives have been considered and tested, not merely rejected.

Simply saying "refraction can't explain this" is not a rebuttal; it is denial. A true rebuttal would show—using measurements and calculations—why refraction is insufficient under the specific conditions being discussed.

I want to repeat this for emphasis: Denial is not a rebuttal. Many arguments gain their persuasive force by acting as if alternatives have already been ruled out when they have not. A proper rebuttal does not merely claim that another explanation is impossible. It shows why that explanation fails.

## Premise

A premise is a foundational statement offered as part of the reasoning of an argument.

Premises are the building blocks from which the conclusion is supposed to follow. They can be observations, definitions, measurements, or other statements assumed to be accurate for the sake of the argument.

For example, "Light travels in straight lines unless acted upon" could function as a premise in some contexts. So could, "Objects hidden below a geometric horizon should not be directly visible under certain measured conditions."

If a premise is false, incomplete, vague, or oversimplified, the argument built on it becomes unreliable as a result. This is one of the most important lessons for the chapters ahead. A conclusion may seem airtight simply because the premises were stated confidently and too quickly. But confidence does not repair an incomplete premise.

## Conclusion

A conclusion is the statement the argument is trying to establish.

It is the final claim that supposedly follows from the premises and reasoning. The conclusion can only be as strong as the premises and the logic connecting them. If either is flawed, the conclusion does not follow, no matter how persuasive it may feel.

That may seem obvious, but in real discussions people often evaluate a conclusion by how strongly it matches their intuitions or how it makes them feel. When reasoning carefully, the right questions are: Does the argument's conclusion actually follow from the support offered? And are the supporting premises true and complete?

## Hidden Assumptions and Incomplete Premises

Now let's think for a moment about our very human tendency to be persuaded by arguments that sound simple and powerful because they leave important information out.

In everyday conversation, people often treat simplicity as a mark of truth. Sometimes that is reasonable. But in many cases simplicity is achieved by omission. A simple argument is often persuasive not because it is complete, but because something important has been left out.

For example, an argument about long-distance visibility may refer to curvature and distance but omit observer height, target height, and the bending of light through air. When that happens, the argument may still feel clean and decisive. Yet its neatness is achieved by leaving important things out.

This is why we must not confuse a simple argument with a strong argument. A strong argument includes all the relevant factors needed to test reality fairly. If essential conditions are left out, then what looks like certainty may only be oversimplification.

This principle will become central in the next chapter. When analyzing arguments, the goal is never merely to ask whether an argument is logical, but to determine whether its premises are complete enough to represent reality accurately.

The parts of an argument defined in this chapter are not usually identified during discussions. However, I define them here so that we can better see whether something important is missing or incomplete in an argument.

**Validity and Soundness**

Two more terms matter if we want to reason carefully.

**Valid Argument**

A valid argument is one in which the conclusion logically follows from the premises if the premises are complete and true.
But validity concerns only the argument's logical structure, not whether its premises are complete or true.

Example:
• All mammals are animals.
• All dogs are mammals.
• Therefore, all dogs are animals.

This argument is valid because the conclusion logically follows from the premises.

However, validity alone does not guarantee a true conclusion. If one of the premises were false or incomplete, the conclusion could be wrong even though the logic is correct.

Another way to say this is that a valid argument is one in which it would be impossible for the premises to be true and complete and the conclusion to be false at the same time.

Validity is only about logical accuracy.

## Sound Argument

A sound argument is a valid argument whose premises are complete and true.
Sound arguments give us reliable knowledge, because they combine correct logic with premises that are both complete and true.

A sound argument does more than feel convincing—its conclusion cannot be false if the premises are complete and true.

Using formal logic, if an argument is valid and its premises are complete and true, then its conclusion cannot be false.
There is no logical space left for error once soundness is established.

This difference between **validity** and **soundness** is crucial for arguments about the shape of the Earth. Many arguments feel airtight because their logical structure is valid. Yet the premises often turn out to be incomplete, overstated, or based on appearances rather than measurements—making the argument valid but unsound.

Logic cannot rescue an argument whose premises do not accurately represent reality.

That is why merely saying "the logic is valid" is never enough. A valid argument can still lead us astray if one or more premises are false, incomplete, or missing relevant conditions.

Soundness is the real goal, because sound arguments are what give us justified conclusions about the world.

This fact is worth repeating: **if an argument is valid and its premises are true, its conclusion cannot be false.**

We should keep this principle in mind whenever we test any argument—not just those about the shape of the Earth.

## When an Argument Feels Stronger Than It Is

One reason some bad arguments are so persuasive is that they combine three things we naturally trust:

- simple wording
- whatever we see
- a valid logical form

That combination can create strong feelings of certainty. A person sees something with his eyes, puts the experience into a simple valid argument, and reaches a firm conclusion. But what we see can be misleading, and validity does not compensate for premises that are based on misleading appearances.

Such arguments can be highly persuasive without being sound.

When debating the shape of the Earth, whether or not arguments are sound is critical because they are often simple, valid, and consistent with what we see, which makes them very convincing even if their premises are incomplete or false. The Earth often appears flat to us. Water appears level. Distant objects can remain visible from much farther away than simple math predicts. Those appearances are all real, but we need to ask if an argument built on them includes all of their possible causes. If not, mere appearances are not enough to justify an argument's conclusion. We should not always believe what appearances seem to imply.

We must learn to separate the persuasive force of an argument from whether it is sound or not, and to restrain our feelings of certainty until all premises have been well tested.

## A Familiar Logical Tool: Modus Tollens

One valid argument form is especially important in discussions about the shape of the Earth because it often feels decisive and final whether or not it is complete or true.

It is called modus tollens, and it has this structure:

- If A, then B
- Not B
- Therefore, not A

When used with accurate premises, modus tollens is powerful. It allows us to reject a model of reality with justified certainty.

But when its premises are incomplete or incorrect, it can produce certainty where none is warranted. The logical structure remains valid, but the conclusion can be wrong because the premises fail to describe reality completely or accurately.

For example, a person may reason:

• If a model were true, we would observe a certain result.
• We do not observe that result.
• Therefore, the model is false.

This reasoning is valid only if the first premise correctly describes what the model actually predicts under the conditions being examined. If the premise leaves out relevant factors, the argument can be logically decisive while failing to be sound.

I want to emphasize this because the most commonly promoted argument against a spherical Earth often relies on this modus tollens structure. It is usually presented in roughly the following way:

• If the Earth were a sphere of a given radius, its surface would curve downward at eight inches per mile squared.
• If that were true, distant objects should be hidden or obstructed by the hump of that curvature.
• But distant objects remain visible.
• Therefore, the Earth is not a sphere.

When first presented, this argument combines mathematics, what we see, and valid logical structure, and it feels powerful, especially when presented by people we regard as intelligent, informed, and trustworthy. Its conclusion can seem indisputable.

In the next chapter, we will examine this argument carefully. We will ask whether its premises are complete enough for the conclusion to follow and whether this argument actually proves what it claims.

## Why This Applies Beyond One Debate

The lesson of this chapter applies far beyond arguments about the shape of Earth.

Whenever we reason about reality, we are tempted to move too quickly from appearances to conclusion. We can mistake what we visually imagine for what is actually true and then turn those imaginations into premises. If the premises are incomplete, the entire argument can look rigorous and definitive while actually resting on our mistake.

That is why scientific models are judged not by a single appearance or a single prediction in isolation, but by their overall explanatory power across many independent systems. A good model must consistently account for measurements made in different contexts by different methods, not just one apparent test.

That broader standard will matter more and more as this book continues. The shape of the Earth is not determined by one sighting, one video, one photograph, or one argument that seems simple. It is tested by multiple systems of measurement that must agree in order to verify that a particular model is correct.

## Common Terms Used Throughout This Book

Now I want to define several additional terms that I use often in this book. Although these words are well-known, I want to be sure you know how I use them, so please do not take offense at me defining what you already know. My goal is clarity, and I believe defining these words will help achieve that.

**Appearance** = When an object is seen or what something looks like from a particular point of view.
**Imaginations** = Mental pictures or ideas about reality.
**Illusions** = Imaginations that misrepresent physical reality and can mislead us about what is physically true.

**Measurement** = A quantified observation used to describe or compare something.
**Tool-obtained measurement** = A quantity of length, distance, or another physical quality obtained using instruments or standardized methods rather than subjective human interpretation.
**Model** = An explanation or representation of something that describes or shows how it works or behaves. A model can be physical, such as a globe (sphere) or flat map, or a mental image.
**System** = A connected set of methods, tools, or processes working together for a particular purpose, such as surveying, navigation, wireless networking, and global positioning.
**Field** = A branch of study or investigation of a particular area of knowledge or reality.
**Verify** = To test or confirm whether something matches reality or agrees with reliable evidence.
**Evidence** = Information, observations, measurements, or facts used to support or test a conclusion.
**Follows** = Logically results from the premises or evidence.
**Believe** = To mentally consider a claim, idea, or conclusion to be true or likely true, whether because of evidence, reasoning, experience, trust, or personal conviction.
**True** = Matching reality or the way things actually are.

## What This Chapter Establishes

This chapter does not ask us to distrust our senses. It asks us to understand what arguments are, how they work, and where they can fail.

It shows that a claim is not established merely because it is clear, intuitive, or repeated confidently. A conclusion must be supported by sound reasoning. That requires more than support; it requires complete premises, justified warrants, appropriate qualifiers, and serious rebuttals to alternatives.

We must therefore remember that the strength of any argument depends not only on its logic but also on the completeness and accuracy of its premises.

Most of all, this chapter establishes a standard we will use throughout the rest of this book:
We should ask not only whether an argument is logical, persuasive, and supported by appearances or assumptions.
We should also ask whether its premises are complete enough and accurate enough for it to be sound and justify its conclusion.

I repeat for emphasis: to avoid being led by an argument to a conclusion that feels certain but is false, we must test both the logic and the completeness and accuracy of its premises to ensure that the conclusion actually follows.

Since people disagree about the shape of the Earth, all arguments in favor of either spherical or non-spherical models should be tested this way. Only valid arguments built on complete and accurate premises justify accepting their conclusions as true.

**This difference between soundness and validity is critical to understand, because if an argument is truly sound, its conclusion is not optional—it must be accepted as true.**

In the chapters ahead, we will apply these concepts to two competing modus tollens arguments about the shape of the Earth and ask which one is sound.

In the next chapter, we will examine the most common "simple math" modus tollens used against a spherical Earth. It is widely repeated because it is valid, it feels mathematically clean, and it seems to be confirmed by what we see with our own eyes, in pictures, and in videos. Our task will be to examine the argument carefully and determine whether its premises are complete and accurate so that the conclusion must follow.

# Chapter 4

## The "Simple Math" Argument Against a Spherical Earth

When analyzing arguments, the goal is not to determine whether they feel convincing, but whether their premises completely and accurately describe reality.

Before criticizing this "simple math" argument against a spherical Earth, let's first think about why so many people have found it persuasive. It does not sound irrational. To the contrary, it sounds like exactly the kind of reasoning careful people should respect: it appeals to mathematics, what we see, and a valid form of reasoning. That is precisely why it deserves to be examined carefully.

The commonly promoted—and most compelling—argument used to reject a spherical Earth that I have heard relies on the familiar logical rule introduced at the end of Chapter 3: modus tollens.

The non-spherical modus tollens argument that prompted me to question my beliefs about the shape of the Earth—which I call the "simple math" modus tollens—is usually presented in roughly the following form:

• If the Earth were a sphere of a given radius, its surface would curve downward at eight inches per mile squared.

• If that were true, distant objects should be hidden or obstructed by the hump of that curvature.
• But distant objects remain visible.
• Therefore, the Earth is not a sphere.

This argument is simple, valid, and seems to be easily testable.

Part of what makes this argument persuasive is that it combines mathematics, what we see, and a valid logical structure.
We trust mathematics we understand. We trust what we see. And we trust easy-to-follow logical rules.

If the premises of this argument are complete and true and its reasoning is valid, then the conclusion must follow with logical certainty and the spherical model must be rejected.

To many, the reasoning feels something like this: if the Earth truly curved at eight inches per mile squared, then distant cities, mountains, or islands should disappear behind the hump of that curvature. Yet we can all see in photographs or videos where such objects still appear at distances that exceed what the "simple math" alone predicts. If those appearances and the "simple math" calculations are considered to be the only factors in determining whether there is curvature or not, the conclusion seems unavoidable: the spherical model must be wrong.

Before evaluating this simple math argument, we will apply the standard established in the previous chapter: the premises must fully account for the specific physical conditions being tested—including those that can produce visual or optical illusions.

As discussed in Chapter 2, optical illusions occur when light is bent, refracted, or otherwise altered before it reaches our retinas. In those cases, our minds form imaginations that do not accurately represent the external world—not because our minds are malfunctioning, but because the light entering our eyes or camera lenses has been physically altered or bent after bouncing off an object.

Light bends downward as it travels through the atmosphere, a phenomenon called atmospheric refraction. This downward bending of light from objects that would otherwise be hidden by curvature allows those objects to be seen by distant observers.

A natural mistake is to interpret such appearances as evidence against curvature when, in reality, they merely demonstrate that light can bend while traveling through the atmosphere and consequently reach our eyes or camera lenses. They are optical illusions, not direct measurements of Earth's geometry.

## The Illusion That Creates False Certainty

The persuasive power of the "simple math" argument comes partly from how human vision works rather than from math or logic—it comes from how the mind interprets what it sees.
It comes from seeing an optical illusion—a false imagination caused by bent light—and our tendency to believe what we see.

When distant objects appear beyond what the simple geometric calculation of eight inches per mile squared allows for, it feels as if we are directly observing a contradiction. The mind interprets long-distance appearances as decisive evidence: *if we can still see it, the Earth cannot be curving as mathematically predicted.*

But that conclusion is logically dependent on an assumption that light has traveled in a straight line from the distant object to the observer, which is not the case.

In reality, the light has been bent by the atmosphere.

Atmospheric refraction brings distant objects into view that would otherwise be hidden by "the hump" of Earth's curvature. A distant object's appearance is not a contradiction of spherical geometry predictions. It is a confirmation that light is bent while traveling through the atmosphere between the object and the observer. It is evidence that temperature layers in the atmosphere—especially over cool water—can bend light enough to offset Earth's curvature and result in powerful optical illusions.

Appearances can feel like visual measurements of curvature, but they are not—they are the result of conditions that the "simple math" argument, in its simplicity, does not include in its premises.

When such appearances are mistaken for direct evidence against curvature, the argument seems to produce certainty—but that certainty is not the result of complete premises; it is the result of confusing optical illusions with measurements of Earth's shape.

Therefore, the "simple math" modus tollens does not fail because its logic is unreliable—it is valid.
It fails because it is built on an incomplete description of reality—one that turns an optical illusion into a foundational premise.

Atmospheric refraction is not an explanation invented after seeing a distant object to defend the spherical model—it is a thoroughly studied physical effect that is routinely measured and accounted for in multiple scientific fields and technological systems.

## What a Properly Constructed Non-spherical Modus Tollens Would Require

Note that in this book, I will often use "non-spherical" to refer broadly to models that reject a spherical Earth, including flat-Earth models and related alternatives.

If a non-spherical modus tollens were going to succeed, it would need to include every factor that affects visibility, not just the "simple math" curvature calculation by itself. An argument with all true premises would have to read something like this:

• If the Earth were a sphere of a given radius, then after accounting for observer height, target height, atmospheric refraction, and the well-known ways light can bend through air, objects beyond a certain distance would be geometrically hidden under all conditions.
• Those objects remain visible even after all of those factors are measured and accounted for.
• Therefore, the Earth is not a sphere.

Notice what this version requires.
It requires measurements rather than appearances alone.

It requires careful consideration of alternative explanations for why objects might remain visible when they ought to be geometrically hidden. It requires demonstrating that refraction is insufficient to alter the expected line of sight and showing that observer height and target height have been included correctly.

It does not simply say, "Those objects remain visible."

It requires testing predicted visibility against calculations that use measurements of Earth's shape and that account for refraction, not merely against what is visible from our perspective.

When these factors are actually measured and included, the argument's conclusion no longer follows from its premises.

## The Geometry Behind the "Simple Math"

Let's look at the basic geometry ("simple math"). For a sphere of radius R (the Earth's mean radius is approximately 6,371 kilometers, or about 3,959 miles), the vertical drop h below a straight tangent line at a horizontal distance d is given approximately by:

$h \approx d^2 / 2R$

In simple terms, this formula tells us how far the Earth's surface falls below a straight tangent line as distance increases.

When converted to common units, this becomes the familiar rule of thumb:

$h \approx 8$ inches per mile$^2$

This is often summarized as "eight inches per mile squared."

So at 10 miles the expected drop is:

$h = 8 \times 10^2$
$= 8 \times 100$
$= 800$ inches
$\approx 66.7$ feet

This calculation is geometrically correct. It describes how far the surface falls below a straight, tangential reference line.

However, this calculation alone does not determine what will or will not be visible through a real atmosphere.

To understand why, we must think about how light actually travels through the atmosphere.

## The Missing Variable in Visibility: Atmospheric Refraction

Light does not travel in perfectly straight lines through air. Air has a refractive index that changes continuously with temperature, pressure, and humidity. Because gravity makes air denser near the surface and progressively less dense with altitude, light rays bend downward as they travel.

Light traveling from a distant city, island, or ship can be bent by the atmosphere and follow Earth's curvature over great distances, allowing the light to reach an observer's eye, telescope, or camera lens.

This bending of light—called atmospheric refraction—is well known and routinely accounted for by professionals in surveying and geodesy. In fact, many claims presented as evidence against a spherical Earth arise from misunderstandings of phenomena like this that have long been thoroughly studied, measured, and accounted for across many scientific fields.

In terms of perception, this produces an optical illusion: the mind correctly interprets the sensory input, but the light itself has been bent along a curved path through the atmosphere. The resulting image can seem to contradict what simple geometry alone would predict, not because the geometric calculation is wrong, but because the light has not traveled in a straight line.

(In surveying, the term surveying refers to the measurement of dimensional relationships—horizontal distances, elevations, directions, and angles—on the Earth's surface, especially for locating property boundaries, construction layout, and mapmaking.

Geodesy is the branch of applied mathematics that determines, through observation and measurement, the shape and dimensions of the Earth and large portions of its surface.)

In these fields, air's refractive index is accounted for through a standard refraction correction. A common approximation treats refraction as reducing the effective curvature by a fraction *k*, where:

$k \approx 0.13$ to $0.20$ (standard conditions)

The effective curvature becomes:

$h_eff = (1 - k)d^2 / 2R$

Under standard conditions with $k = 0.13$, about 13% of the geometric drop is canceled by refraction.

But atmospheric conditions are not always standard.

## When Refraction Becomes Stronger Than Curvature

Over water—especially cold water with warmer air above it—temperature gradients can become steep. This produces super-refraction, where light bends more strongly downward than usual.

In extreme cases, refraction can bend light more than the Earth curves, producing an apparent and memorable lifting of distant objects above the horizon.

In those cases, distant objects may appear:

- fully visible when geometry alone predicts partial or complete obstruction
- vertically compressed or stretched
- or even elevated above the apparent horizon, as if floating.

This is not hypothetical. It is measured, modeled, and predicted.

Because refraction is so well understood, it is not only something we explain after the fact—it is something we can predict in advance. When atmospheric conditions such as temperature gradients, pressure, and humidity are measured, scientists and engineers can anticipate when distant objects will appear more visible than usual, or even appear elevated above the horizon. In other words, these observations are not unexpected—they are predicted by the same physics that explains them.

The key thing to understand is that atmospheric refraction allows light to follow the Earth's curvature more closely than simple geometry alone predicts, making distant objects visible beyond what a straight-line path of light would allow.

Mathematically, this happens when the refraction coefficient $k$ exceeds the standard value.

If:

$k > 0.5$

then refraction bends light downward more strongly than the surface curves away. Under those conditions, distant objects can appear higher than expected, not lower.

## Why Visibility Cannot Disprove Curvature

Seeing objects at a distance is not a direct measure of geometry; it is simply seeing light that has bounced off the objects, nothing more. We know for certain that what we are seeing can be an optical illusion.

This brings us back to the original argument against the spherical Earth model, stated roughly as:

"If the Earth were a sphere of a given radius, its surface would curve downward with distance, creating a hump that would obstruct the visibility of objects beyond an easily calculated distance. But we can still see objects beyond that distance. Therefore, the Earth is not a sphere."

That version of modus tollens (unsound because its premises are not complete) treats visibility as if it were a direct test of curvature.

But visibility depends on:

- observer height
- target height
- Earth's curvature
- and, differing with weather, the bending of light through air.

Because each of these variables can be measured, the resulting visibility is not a mystery. Under known atmospheric conditions, the bending of light can be calculated, and the visibility of distant objects can be anticipated before any observation is made. This means that seeing farther than simple geometric expectations is not evidence against curvature—it is evidence that all variables, including refraction, must be included.

When distant objects remain visible beyond the geometric horizon—or even appear elevated—it does not mean that curvature is absent.

The underlying geometric relationships remain unchanged.
And the curvature remains unchanged.

It simply means the light reached our eyes or zoom lenses—and produced the appearance of the distant object—as a result of atmospheric refraction bending its path through the air.

## Applying the Reasoning: Lake Michigan and Hawaii

With these considerations in mind, we can now examine two real-world examples often cited in support of the argument: the Chicago skyline visible across Lake Michigan and the visibility between some of the Hawaiian islands.

These cases feel especially powerful because we naturally trust what we see. We can point to videos, photographs, and firsthand observations. But seeing something is not the same as measuring it—and that is why visual appearances must be tested against measurements. Optical illusions can feel so convincing that measurements are often skipped.

Let's think about what happens over Lake Michigan.

## Lake Michigan: A Real-World Example

The distance from the Michigan shoreline near St. Joseph to the Chicago skyline is roughly 60 miles.

Using the geometric curvature formula:

$h \approx 8$ inches per mile$^2$

Because the distance is squared (for example: 10 miles times 10), the curvature drop increases rapidly as the distance grows.

The expected geometric drop over 60 miles is:

$$
\begin{aligned}
h &= 8 \times 60^2 \\
&= 8 \times 3600 \\
&= 28{,}800 \text{ inches} \\
&\approx 2{,}400 \text{ feet}
\end{aligned}
$$

Let's pause and think about what this number means.

That simple math—pure geometry—tells us that the surface of the lake curves downward by about 2,400 feet—nearly half a mile—relative to a straight, tangential line across sixty miles of water.

That number follows directly from the geometry and represents a significant drop over that distance.

However, this calculation alone does not determine what will or will not be visible through a real atmosphere.

We must also add the real conditions of Lake Michigan. The lake is often cold, especially in spring and early summer. The air above it is frequently warmer. This creates a strong vertical temperature gradient: dense cooler air near the surface and lower-density warmer air higher up.

Under these conditions, light rays bend downward more strongly than under standard atmospheric conditions.

As noted earlier, these conditions are not guesswork. They can be measured, modeled, and used to predict when visibility will exceed simple geometric expectations. When observers report unusually clear or elevated views of distant skylines, those observations often coincide with precisely the atmospheric conditions known to produce stronger refraction.

In atmospheric refraction, the refraction coefficient $k$, which is typically around 0.13 under standard conditions, can increase significantly. In cases of strong over-water refraction, values of $k$ can approach or even exceed 0.25–0.40, meaning that roughly a quarter to nearly half of the geometric curvature can be offset by the downward bending of light in the atmosphere.

When observer height and target height are included—along with refraction—it is not surprising that portions of the Chicago skyline, even sixty miles away, can remain visible, appear compressed, distorted, or even lifted above the visible horizon.

When we "see over the curve," it is the result of light following a bent path through the atmosphere before reaching our eyes or the lenses of cameras.

## Why Hawaii Works the Same Way

The same reasoning applies on an even larger scale in the Hawaiian Islands.

The distance between Kauai and Oahu is about 108 miles.

The geometric curvature drop below a straight tangential line across that distance is enormous—well over 7,000 feet, more than a mile of curvature drop across the ocean between them.
Yet under the right atmospheric conditions, portions of those islands can sometimes remain visible.

As before, when refraction, observer height, and target elevation are measured and included, partial visibility is not only possible—it is expected under the right conditions. Those conditions can be identified in advance, allowing scientists to predict when such visibility will occur.

These cases do not falsify curvature. They demonstrate how light and curvature interact.

Therefore, in all cases of seeing what the "simple math" says we should not be able to see, what we should pay attention to is not whether something appears, but whether all variables affecting visibility have been measured and accounted for.

## Why Zoom Lenses and Telescopes Seem to "Bring Boats Back"

This same misunderstanding appears in videos of boats disappearing over the horizon and then "reappearing" when zoom lenses are applied.

When a boat moves far enough away, the lower portions of it disappear first. To the naked eye, the boat seems to sink.

Zoom lenses simply magnify the light that is already reaching them; they do not extend the distance the camera can see.

When a boat is near the limit of visibility, refraction may still bend some of its light toward the observer. Optical zoom lenses can capture and enlarge that bent light.

But there is a limit to how far light can bend. Once a boat moves beyond the distance that refraction can cause the light coming from the boat to reach the observer, no amount of zoom can recover it.

At that point, the curved surface of the Earth is preventing the light reflected from the boat from reaching the observer.

This tells us something important: the visibility of distant objects is determined by geometry and refraction, not by geometry alone.

## Measurement Versus Imagination

Light's bendability is why we cannot rely on simple curvature calculations and appearances alone to decide the shape of the Earth.

The Earth may appear flat from our perspective, but measurements tell a more complex story.

In the pages ahead, we will examine how real-world measurement systems account for refraction—and how their measurements consistently agree with a spherical Earth.

## What the Logic Finally Tells Us

The modus tollens form used in the simple math argument of the previous chapter is logically valid. The problem is in the premises.

When visibility alone is treated as a decisive test of curvature, important physical factors are being left out.

The problem is not the logical form. The problem is that the premises are incomplete.

## What This Chapter Establishes

This chapter does not ask us to distrust our senses. It asks us to examine them with careful reasoning.

Seeing beyond a geometric horizon is not evidence that Earth is non-spherical. It is evidence that light can bend downward as it travels through the atmosphere. What gives the "simple math" argument its persuasive force is that this effect creates a convincing optical illusion—one that feels like a direct measurement, but is not.

Completing incomplete premises is not a minor adjustment—it changes the entire argument and may change the conclusion. When the premises change, the conclusion must change accordingly.

Claims that seem to contradict the spherical model can appear strong when they leave out well-tested and established physical behaviors, such as atmospheric refraction and optical illusions.

In the pursuit of truth, we should all ask whether a different modus tollens—built on complete premises and tested against measured reality—leads us to a conclusion that actually corresponds to reality. The reason is that conclusions are not made true because they feel persuasive or intuitive to us. They are made true by logically following from complete premises that accurately describe reality.

# Chapter 5

## A Measurements-Based Modus Tollens

In the previous chapter, we examined the most compelling argument commonly used to reject a spherical Earth: a "simple math" modus tollens built on spherical geometry calculations and visual observation. We saw why it feels decisive—and why it cannot be sound unless observer height, target height, and atmospheric refraction are measured and included.

The argument fails not because its logical form is invalid, but because its premises are incomplete.

In this chapter we will examine a different modus tollens built on measurable premises.

This measurements-based form of modus tollens does not rely on appearances, perspectives, or visibility alone. Instead, it relies on measurements that can be tested, repeated, and compared across multiple independent fields and systems.

Under careful scrutiny, this modus tollens argument does not weaken—it grows stronger. And if its premises are shown to be true, then its conclusion does not merely seem persuasive—it becomes logically unavoidable.

The task of the next several chapters is to test whether this modus tollens argument is sound.

Our goal is not to win an argument, but to understand why one version of modus tollens fails while another succeeds.

## Why Modus Tollens Is Not the Problem

It is important to emphasize that modus tollens itself is not the issue. Modus tollens is a valid form of reasoning:

- If A, then B.
- Not B.
- Therefore, not A.

When its premises are true, it is one of the strongest tools we have for rejecting false models of reality. When its premises are incomplete or false, however, it can feel decisive while leading us to the wrong conclusion.

The "simple math" modus tollens for a non-spherical Earth that we examined earlier failed because it treated visibility as a purely geometric matter—mistaking an optical illusion for a direct measurement—and ignored measurable variables that affect how light is bent.

So now we can ask:

Can a modus tollens be constructed using premises grounded not in appearances, but in independent, cross-checked measurement?

## Agreement as Evidence

The spherical Earth model is not based on a single observation or a single calculation. It is a geometric model with specific quantitative consequences. If the Earth were not spherical—if it were flat or some other non-spherical shape—then measurement systems built on spherical geometry would fail to agree with one another.

But as you will see, they do agree. And that agreement is not trivial.

When independent measurement systems—developed for different purposes, using different instruments, by different communities of people—all agree mathematically only when spherical curvature is included, that agreement itself becomes powerful evidence of a spherical model of the Earth.

The engineers who design navigation systems, communication networks, and positioning technologies are not attempting to defend a philosophical model of the Earth. Their goal is simply to build systems that function reliably.

The spherical model persists because it allows those systems to function reliably and consistently explains the measurements on which they depend.

A model can survive rhetorical attack for years, but no model survives engineering failure for long. A model earns confidence not by persuasion, but by surviving repeated tests against measurements.

This is not an appeal to authority. It is an appeal to measurements that professionals in multiple fields verify every day—measurements that independently, repeatedly, and in practice agree on spherical geometry.

## A Measurements-Based Modus Tollens

Let's now state the argument carefully.

### Premise 1 — Conditional

If the Earth were not spherical (that is, if it were flat or some other non-spherical shape), then independent measurement systems that rely on spherical geometry would fail to agree mathematically with one another.

These systems include:
• large-scale land surveying and geodesy (the mathematics of the size and shape of the Earth)
• long-distance navigation and route planning
• radio-signal propagation and horizon calculations
• satellite tracking and orbital mechanics
• Global Positioning System (GPS) calculations

But these systems, which are examined in this book, are not an exhaustive list.

They are merely a representative sample.

The evidence examined in this book represents only a small and accessible portion of a much larger body of interconnected and well-verified knowledge supporting a spherical Earth.

Each of these systems measures different physical quantities using different tools and methods. Their agreement is not assumed; it is tested constantly in real-world practice.

The claim is not that these systems assume a sphere, but that they only function when their calculations are based on real-world measurements indicating spherical geometry, and that they simply would not work if those calculations were based on any non-spherical Earth model.

Accepting a non-spherical Earth is not a small adjustment to a single belief. It is more like replacing an entire belief system with another. It requires reinterpreting or rejecting a vast network of measurements developed across many independent fields of science and technology over centuries. Surveying, navigation, radio engineering, orbital mechanics, and positioning systems—all discussed in this book—rely on mathematical relationships that agree with one another. To replace the spherical model, an alternative must explain not only one set of observations, but also account for many interwoven measurement systems that agree.

Changing our conclusions about the shape of the Earth therefore requires more than rejecting a single idea. It requires adopting a new model that can make sense of those interconnected measurements as well as—or better than—the spherical model does.

The chapters that follow will examine several of these systems in detail so that we can see how their mathematical results agree when Earth is modeled as a sphere.

If the Earth were not spherical, their equations would not reconcile. Their corrections would introduce error rather than eliminate it. Their predictions would drift apart rather than come together.

**Premise 2 — Negation of the Consequent**

But these independent measurement systems—surveying, navigation, radio propagation, and global positioning—developed for different purposes do agree mathematically, consistently, predictably, and precisely when Earth is modeled as a sphere with a specific, measurable radius.

When curvature is included:

• surveying triangles exhibit predictable spherical excess
• navigation routes match great-circle calculations rather than flat-plane predictions
• radio horizons extend in proportion to antenna height exactly as curved-surface geometry predicts
• satellites maintain stable, measurable orbits at calculated altitudes, making GPS possible
• GPS receivers determine position accurately only when curvature and relativistic corrections are included

When curvature is excluded from the calculations, these systems do not reconcile. The numbers fail to agree. Predictions degrade.

In other words, when the curvature corrections that normally stabilize these systems are omitted, errors accumulate and system performance breaks down.

**Conclusion — Modus Tollens**

*Therefore, the Earth cannot be flat or any other non-spherical shape.*

In practice, the only geometric model that allows these independent systems to reconcile is a spherical Earth (more precisely, an oblate spheroid). Rejecting non-spherical alternatives leaves spherical geometry as the only model that consistently fits the cross-system measurements.

## Why This Argument Is Valid

The structure of the reasoning is straightforward:

- If A, then B.
- Not B.
- Therefore, not A.

The conclusion follows from the premises. There is no structural flaw.

The non-spherical modus tollens in the last chapter was also valid in form. Its weakness was not logic—it was incomplete premises. The same test must be applied here.

## Why This Argument Is Sound

Soundness requires both a valid structure and true premises.

The crucial difference between this argument and the earlier one lies in how the premises are established.

The non-spherical modus tollens relies on visual appearances interpreted as if geometry alone determines visibility.

By contrast, the measurements-based modus tollens relies on premises grounded in repeatable, tool-obtained measurements made across multiple independent systems. These premises are not based on appearances alone or on subjective interpretations; they are based on numerical measurements obtained by instruments and verified by practical success.

Surveyors do not "see" curvature. They measure angles and distances and discover that two-dimensional, planar geometry fails while three-dimensional, spherical geometry succeeds.

Navigators do not imagine curvature. They calculate routes and discover that great-circle paths work while flat-plane routes accumulate error.

Radio engineers do not infer curvature from appearances. They measure signal horizons and find that range increases precisely with antenna height as curved-surface geometry predicts.

GPS systems do not assume curvature out of tradition. They would not function accurately without curvature and relativistic corrections in their equations.

These measurements are not isolated—they come together in agreement.

And that agreement is the key evidence supporting Premise 2.

## How Many Measurements Must Agree?

Any single measurement can be misleading. Any single instrument can malfunction. Any single observation can be misinterpreted. But this argument is not based on one measurement or even on only a few. It is based on countless independent measurements that mathematically agree with one another.

In Chapters 7 through 10, I will describe how four independent measurement systems rely on billions of measurements and on spherical geometry in order to function accurately.

As you read the rest of this book, I ask you to think about two common quotes in conspiracy communities: "How many coincidences before it becomes mathematically impossible?" and "How many coincidences do you need before you believe?"

This argument is not about coincidences. It is about measurements. Every day, countless independent measurements are made in surveying, navigation, radio propagation, and global positioning systems that all repeatedly agree mathematically when Earth is modeled as a sphere with a specific, measurable radius.

Since these systems use spherical geometry in their calculations, the spherical Earth model best explains the consistent and precise agreement across all of them.

As you learn how these systems work, I hope you will ask yourself: how many independent measurements, repeated across how many independent systems, must agree before it becomes unreasonable to disbelieve what they consistently indicate—that the shape of the Earth is spherical?

## Openness to Falsification

Another important difference between the two arguments presented in this book is how they respond to challenge.

The non-spherical modus tollens often becomes psychologically one-directional. Once someone believes curvature has been visually disproven, all contrary evidence is rejected as fabricated, manipulated, or conspiratorial.

The measurements-based spherical Earth modus tollens remains open to falsification.

If surveying networks stopped showing spherical excess,
if navigation routes stopped matching great-circle predictions,
if radio horizons stopped scaling with antenna height,
if satellites failed to orbit as calculated,
if GPS ceased to function without curvature corrections—

then the premises of this argument would be false, and the spherical Earth conclusion would have to be reconsidered.

That openness to falsification is not weakness.

It is the defining strength of sound reasoning.

The remaining chapters test these premises by examining four independent measurement systems on which many modern technologies depend. Each system measures different physical quantities using different instruments, yet their calculations agree only when Earth is modeled using spherical geometry with a specific measurable radius.

Different measurements.
Same result.

## What This Chapter Establishes

This chapter does not ask us to reject what our senses tell us. It asks us to place that information in proper context.

It does not ask us to trust authority. It asks us to examine measurement systems built on instruments and mathematics that have been tested, corrected, and refined by millions of people across centuries—systems that agree mathematically only when spherical geometry is used.

The “simple math” modus tollens for a non-spherical Earth fails because its premises are incomplete.

The measurements-based modus tollens for a spherical Earth succeeds because its premises are grounded in cross-system measurement agreement.

When logic is built on accurate premises, it does not lead us away from reality.

It leads us toward it.

In the chapters that follow, we will examine these independent measurement systems one by one. Surveying measures angles across large regions. Navigation measures long-distance routes. Radio engineering measures horizons. GPS measures position using orbiting satellites.

Each field uses different instruments and measures different physical quantities.

Yet all arrive at the same result: Earth behaves as a sphere with a measurable radius of about 6,371 kilometers (3,959 miles). In scientific and engineering fields, saying Earth "behaves as a sphere" means that independent measurements and technological systems consistently agree with the predictions made by spherical geometry.

That same result is not accidental. As we asked earlier: how many independent measurements, repeated across how many independent systems, must agree before it becomes unreasonable to disbelieve what they consistently indicate—that the shape of the Earth is spherical?

Now let's pause the argument for a chapter and think about something that influences how we evaluate evidence itself.

In the next chapter, I will explain the intellectual discipline we should practice before agreeing with a conclusion.

The way we mentally prepare to examine evidence often determines what we are willing to recognize when we see it.

# Chapter 6

## Intellectual Discipline Before Drawing Conclusions

Before agreeing with any conclusion—including the one just presented—we must first examine how we approach evidence. Intellectual discipline is not about suppressing doubt or protecting beliefs. It is about cultivating the humility and fairness required to understand competing explanations before deciding which one best accounts for the evidence—and therefore deserves our confidence.

The next several chapters include assignments designed to help you examine these ideas for yourself. If you skip them because you already feel confident in your conclusion, then your belief is not being tested—it is being protected from challenge.

One of the greatest obstacles to intellectual discipline is confirmation bias—the tendency to notice, emphasize, and remember evidence that supports what we already believe, while overlooking or minimizing evidence that challenges us. This tendency is a universal feature of human thinking. Unless we consciously guard against it, we will probably fall prey to it.

Up to this point, we have discussed how arguments work. We have examined premises, warrants, rebuttals, and logical form. We have seen that arguments can be powerful—but they only produce true conclusions when their premises are complete and accurate. The premises of the "simple math" modus tollens proved incomplete.

In Chapter 5, we examined a different argument: the measurements-based modus tollens. Its conclusion depends on a key premise—that independent measurement systems operating in the real world agree mathematically when Earth is modeled as a sphere with a specific measurable radius.

Before we examine that premise, we must decide how we will approach the evidence, because our approach often determines what we see and what we miss.

## The Danger of Confirmation Bias

**Confirmation bias** is our natural human tendency to favor evidence that reinforces ideas we already believe while resisting evidence that challenges our beliefs. It affects everyone—scientists, skeptics, believers, and critics alike. It is a subconscious and automatic tendency, and it becomes stronger the more closely our beliefs become connected to our identity, certainty, reputation, or relationships with other people. Any evidence that challenges our beliefs can feel threatening to everything connected to them, and feeling that threat can tempt us to defend our beliefs rather than fairly evaluate what the evidence implies.

Once we feel certain the Earth is either spherical or non-spherical—especially after accepting an argument we believe rules out alternatives—it becomes difficult to fairly consider other possibilities. From that point forward, we instinctively look for explanations that preserve our beliefs, give greater attention to evidence that confirms them, and overlook evidence against them.

If we do not intentionally examine evidence that supports the model we currently reject, we may end up seriously examining only one side. And if we do that, we are no longer fairly evaluating the argument—we are selecting the conclusion in advance. Guarding against that requires a simple but difficult step: being willing to say to ourselves, "Maybe my beliefs are wrong," and then verifying that our premises are complete and accurate.

## When Distrust Becomes the Filter

Many readers of this book—including myself—do not begin from a place of trust in institutions. We have seen enough deception and manipulation to know that conspiracies can be real, which often makes us quicker to question official explanations—and sometimes for good reason.

But let's acknowledge something about distrust:

Distrust can be a starting point for investigation—or it can become a filter that determines conclusions in advance.

When distrust is used as a starting point, it leads us to ask: "How does this actually work? Can it be tested? Where might this claim fail?"

But when distrust becomes the filter, we tend to do something else.

We reject explanations before they are understood.
We reject measurements before they are examined.
We treat entire systems—surveying, navigation, radio engineering, and global positioning—as deceptive in principle rather than evaluating them in detail.

When we do that, the issue is no longer the strength of the evidence. Instead, distrust of the source becomes a reason to reject evidence regardless of its content—and soon we may begin dismissing any evidence that challenges our beliefs without fairly considering it.

That is belief protection.

A claim that multiple independent systems—developed by different people, in different fields, for different purposes—are all functioning as part of a coordinated deception or conspiracy is not a small claim. It is a comprehensive one.

And like any other claim, it must itself be tested—not assumed.

Intellectual discipline requires something very difficult:
It requires examining how these systems actually work before rejecting them as untrustworthy.
It requires allowing our beliefs about the shape of the Earth to be challenged by how real-world technologies function.
It requires being willing to say to ourselves, "Maybe this claim is true—and maybe my beliefs are wrong."
It requires allowing the possibility—however uncomfortable—that a source we distrust may still provide evidence that challenges our beliefs—and may require us to revise them.
It requires holding both models of the Earth in mind at the same time—without deciding in advance which is correct—rather than treating one as possible and the other as impossible during investigation.

The goal is not to abandon distrust of the mainstream narrative, but to understand what these systems claim to do before rejecting them as untrustworthy—and then decide.

In the chapters that follow, you will have the opportunity to understand how surveying, navigation, radio engineering, and global positioning systems work—and then decide for yourself which explanation makes the most sense of the measurements these systems provide.

We do not need to determine whether people can be trusted, but whether the systems themselves—when understood and tested—behave in ways better explained by one model or the other.

We cannot know in advance whether one explanation is better than another—the systems must first be understood and tested.

## Understanding Both Models

If you are familiar with Flat Earth Dave's material available on his website, you know that it attempts to explain virtually every appearance according to a non-spherical model. It provides alternative interpretations for sunsets, horizons, long-distance visibility, flight paths, eclipses, and more. Many find it compelling because it offers a complete interpretive framework.

I encourage you to examine those explanations. Here is the discipline I am asking you to practice:

As you read the next four chapters, first seek to understand how the spherical model explains these systems and the measurements they produce on their own terms.

Then, go to flatearthdave.com and other flat-Earth resources, such as The Flat Earth Society and its wiki. Study how those sources explain how these systems work, and then compare those explanations with those of the spherical model.

Ask yourself:

- Which model accounts for more of what is measured across independent systems?
- Which model requires fewer added assumptions and explanations to reconcile observations?
- Which model predicts measured outcomes before they occur?
- Which model remains consistent with measurements across independent systems?

Do not give one model a full hearing while rejecting the other in advance as impossible. We must be willing, at least temporarily, to recognize that our current beliefs may be wrong.

Intellectual honesty requires understanding both arguments fairly before drawing conclusions.

## Humility in the Face of Evidence

Humility does not mean indecision.
It means refusing to protect a conclusion before it has earned it.
It means being willing to say to one's self, "I might not fully understand the other side's claims."
It means trying to understand how experts on both sides explain multiple interacting lines of evidence for and against each model.
It means reconsidering conclusions that feel settled whenever the evidence begins to challenge them—and being vigilant not to allow emotional attachment to a belief, or statements and quotations taken out of context, to substitute for evidence.

In the next chapters, we will examine measurement systems that operate independently. As you examine the evidence those systems provide, resist the urge to defend your beliefs too quickly.

If non-spherical models explain the evidence better, that will become clear.
If the spherical model explains it better, that will become clear.
Mental humility will then follow the evidence where it leads.

## Homework: Testing the Premises

To help you practice that humility and avoid confirmation bias, each of the next four chapters includes a homework assignment—on Surveying, Navigation, Radio Communication, and GPS.

Their purpose is simple: to help you understand how the spherical Earth model explains the measurements we are discussing.

You may have heard the phrase, "Do your own research." These assignments are designed to help you do exactly that.

Watch experts explain how these systems work. Think about whether their explanations are coherent and agree across independent systems.

## How the Next Chapters Test the Arguments

In Chapter 5 we stated a measurements-based modus tollens whose conclusion is that the Earth cannot be flat or any other non-spherical shape. That argument depends on a key premise: that independent measurement systems agree mathematically when Earth is modeled as a sphere with a specific measurable radius.

The next four chapters examine that premise directly. Each chapter focuses on a different measurement system used in the real world: surveying, navigation, radio communication, and satellite-based positioning. These systems were developed independently, for different practical purposes, using different instruments and methods.

If these systems agree only when spherical geometry is used, then the premise is supported. If they do not agree, the argument fails.

Your task in the chapters that follow is therefore simple: examine how each system works and decide whether the measurements truly agree about the shape of the Earth.

Remember that our goal is not simply to answer a scientific question, but to understand how good reasoning distinguishes sound arguments from persuasive ones. We therefore need to test the premise presented in the measurements-based modus tollens—that multiple independent methods of measuring our world really do *"mathematically agree—consistently, predictably, and precisely—when Earth is modeled as a sphere with a specific measurable radius."*

If their agreement holds and the premises are true, then the argument is sound.

Once you have honestly examined how both models explain the same measurements, you will be ready to make a fully informed decision about which model best accounts for the evidence and deserves your confidence.

## Beyond This Debate: A General Method

The discipline practiced in this chapter extends beyond the question of Earth's shape. Every argument—on any subject—depends on premises that must be both accurate and complete. When premises are assumed to be accurate and complete rather than tested, conclusions can feel convincing while still being wrong. For that reason, the habit of carefully examining premises is not limited to this topic. It is a skill that applies to any claim we encounter—especially those that are widely debated or emotionally charged. As you work through the next chapters, the goal is not only to evaluate one argument, but to practice a way of thinking: to question, to verify, and to ensure that arguments are not accepted merely because they are persuasive, but because their premises have been tested and found to be both accurate and complete.

The next four chapters will examine several measurement systems so that you can cross-check the evidence for yourself and determine whether the premise—that independent measurement systems mathematically agree when Earth is modeled as a sphere—is accurate and complete.

The goal is not to confirm what you already believe, but to discover which explanation best accounts for well-established measurements across independent systems. That discovery is only possible if you remain willing, throughout the process, to say to yourself, "Maybe my beliefs are wrong." The willingness to seriously consider that possibility is one of the strongest protections against confirmation bias.

# Chapter 7

## Surveying: Measuring Earth's Curvature Through Angles

Human beings naturally form conclusions about the world through perspective. The horizon looks flat. The ground feels level. From where we stand, the Earth appears to be a flat, non-rotating, stationary plane.

But the technological systems we rely on to measure the world do not operate through perspective. Surveying instruments do not rely on appearances or intuitive judgment; they measure quantities rather than interpreting appearances. They operate through measured angles, measured distances, and mathematical relationships that must agree for the system to function reliably.

Surveying instruments do not care what the Earth looks like. They enable precise measurements. If the spherical geometry used in their calculations were wrong, the systems built upon those calculations would fail.

Surveying provides one of the clearest and oldest examples of this principle.

Surveyors do not rely on what the land appears to do. They measure angles, distances, and the direction of gravity. If the Earth were not a sphere, surveying across large distances would obey flat (Euclidean) geometry. If the Earth were curved, those measurements would reveal curvature in their angular relationships across large distances. And that curvature is what the measurements consistently reveal.

## Why Surveying Exists

Surveying exists for practical reasons. Roads must meet. Bridges and pipelines built from opposite directions must align. Tunnels drilled from opposite sides of mountains must connect. Large property boundaries must close precisely. Surveying guides all such construction projects so they accomplish exactly what is planned.

If surveyors used the wrong geometric model of Earth's shape, these projects would not "almost work." Errors would accumulate; structures would miss by measurable amounts. Instead, they align—consistently, worldwide.

That reliability comes from using the correct geometric model of Earth's shape—a spherical one.

## What "Level" Actually Means

At the heart of surveying is precise angle measurements using instruments such as theodolites and total stations, which you will see in the homework videos below. Before measuring any angles, the instruments are leveled.

This step is crucial.

To a surveyor, "level" does not mean "parallel to what appears flat." It means perpendicular to the direction of gravity at that location—inward toward the center of the Earth. Since gravity points inward everywhere, each location has its own slightly different level direction—and therefore its own tangent plane.

Surveying instruments are leveled using gravity itself—traditionally with plumb lines and bubble levels, and today often with electronic tilt sensors. When an instrument is leveled, its horizontal plane is tangent to the Earth at that point.

Because gravity points toward the Earth's center rather than in parallel directions, each leveled instrument establishes a slightly different tangent plane.

Over short distances, this difference is negligible.
Over large distances, it accumulates—and that accumulation becomes measurable.

## Measuring a Large Triangle

Imagine three high, fixed points—A, B, and C—perhaps mountaintops, survey towers, or permanently established geodetic markers separated by tens or even hundreds of miles. Across the world, surveyors rely on networks of such fixed reference points—carefully measured and maintained over time—to build consistent, large-scale maps and infrastructure.

From point A, a surveyor measures the angle between lines of sight to B and C.
From point B, the angle between A and C is measured.
From point C, the angle between A and B is measured.

These three measured angles define a triangle traced across the surface of the Earth.

On a flat surface, the interior angles of any triangle must add up to exactly:

**180°**

This is a fundamental result of flat (Euclidean) geometry.

But when surveyors measure triangles that span sufficiently large areas—using these fixed reference networks—the angles do not add up to exactly 180 degrees.

They add up to more.

## The Spherical Excess

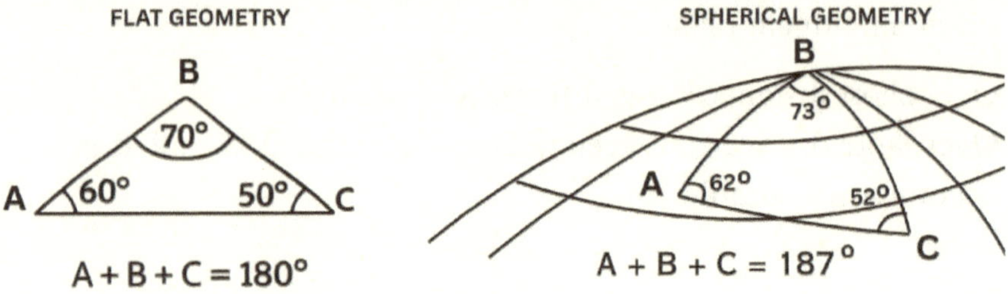


The amount by which the sum exceeds 180 degrees is called the spherical excess.

If the three measured angles add up to:

$180° + E$

then E is the spherical excess.

This excess is not random. It follows a precise mathematical relationship predicted by spherical geometry:

$E \approx A / R^2$

where:

- A is the area of the triangle
- R is the radius of the Earth
- E is measured in radians (or converted to degrees or arc-seconds)

The larger the triangle's area, the larger the excess. And most importantly, the excess increases predictably with area exactly as spherical geometry predicts.

This is not a visual phenomenon, nor is it inferred from photographs. It is measured directly from the angle readings at each corner of the triangle and is predicted precisely using the above equation when the radius of the spherical Earth is used.

## A Worked Illustration

To see how this works in practice, imagine a surveying triangle covering 10,000 square kilometers.

Earth's average radius is approximately

$R \approx 6{,}371$ km ($\approx 3{,}959$ miles)

Then

$R^2 \approx 40{,}600{,}000$ km$^2$ ($\approx 15{,}680{,}000$ square miles)

Applying the spherical excess formula to a triangle with an area of about 10,000 km$^2$ ($\approx 3{,}860$ square miles)

$E \approx 10{,}000 / 40{,}600{,}000$

$\approx 0.000246$ radians

Converting to degrees

$\approx 0.014°$

This corresponds to about 50 arcseconds.

Fifty arcseconds is far too small to see with the eye—but easily detectable with precision instruments. When surveyors measure triangles of that scale, they consistently detect excess values that scale with the area exactly as the spherical formula predicts.

If the Earth were flat, the angle sum would remain 180 degrees regardless of triangle size.

It does not.

## Why Refraction Does Not Eliminate the Effect

Surveyors are well aware that light bends in air. That is why they:

- take reciprocal measurements from opposite directions
- average repeated observations

• measure under stable atmospheric conditions
• apply known refraction corrections when necessary

Refraction can disturb individual measurements, but it does not manufacture spherical excess across large geodetic networks.

Spherical excess persists after corrections. It grows with area. It matches spherical geometry.

Perspective and visual illusion cannot produce this effect. Only spherical geometry can.

## Why This Matters Practically

Surveying is not theoretical. It governs how real infrastructure is built.

Long tunnels drilled from opposite ends meet with astonishing precision. Large-scale mapping systems close properly. Continental geodetic networks reconcile mathematically.

If curvature corrections were being applied to a non-spherical Earth, errors would accumulate instead of cancel. Instead, applying spherical geometry prevents error.

The measurements agree.

## What Surveying Shows

Surveyors do not see the Earth curve. Their eyes report the same flat appearances everyone else experiences.

But measurements and appearances disagree.

Surveying does not rely on appearances.
It relies on repeated angle measurements across large distances.

And when those measurements are compared across many independent surveys, they consistently agree with spherical geometry.

When large triangles are measured:

- their angles add up to more than 180 degrees
- the excess increases predictably with area
- the magnitude corresponds to a sphere with a radius of about 6,371 kilometers (≈ 3,959 miles)

If the Earth were not spherical, these independent measurements would not consistently agree.

Surveying therefore provides a clear measurements-based demonstration of spherical geometry—one independent of vision, perspective, or photographic interpretation.

And surveying is only one system in which spherical geometry appears.

Modern satellite-based systems like GPS extend the global network of fixed survey reference points into space, allowing positions to be measured with even greater precision.

Although the Earth may appear flat at human scale, careful surveying measurements suggest that its surface is consistent with that of a sphere.

Surveying reveals curvature through measured angles; navigation reveals the same geometry through the paths that successfully carry us across the Earth.

In the next chapter, we will examine navigation, where long-distance routes succeed only when calculated along great-circle paths.

# Homework

## Understand Surveying According to Both Models

### Watch: Spherical Model

"How does land surveying work?" by: Practical Engineering youtube.com/watch?v=SPCewaAfqPA

"Surveying Chapter 4: Curvature & Refraction" by: Astrid Northrup youtube.com/watch?v=BTVYpNYvxt0

What are Geodetic Datums? By: The COMET Program/MetEd youtube.com/watch?v=kXTHaMY3cVk

"Curvature and Refraction" by: EngineerItProgram youtube.com/watch?v=CFb8FO9DoLU

"The math that proves the Earth isn't flat" by: Zach Star youtube.com/watch?v=tejB8pFOYD8

"Why there are no Flat Earth Surveyors" by: Dave McKeegan youtube.com/watch?v=iXYPKfhCkRU

"Flat Earthers butcher 'Level'" by: Dave McKeegan youtube.com/watch?v=II5V39oohck

You may also search YouTube for:
- "Spherical excess explained"
- "How surveyors account for Earth's curvature"

Then, deliberately seek out flat-Earth explanations of how spherical excess and large-scale surveying are interpreted within a non-spherical model. Compare those explanations directly with what surveyors measure.

## Investigate

As you watch, ask:

- What angles are surveyors actually measuring?
- What angle sum would a flat surface predict?
- What angle sum do surveyors actually measure over large areas?
- Why do the angles inside large surveying triangles add up to more than they would on a flat surface?
- How do these videos about surveying suggest a shape of the Earth, directly or implied?

Your task is not simply to watch, but to understand and compare why the angles inside a triangle measured from each corner add up to a number greater than they would if the Earth were flat.

## Reflect

Ask yourself:

- Can I restate the explanation fairly in my own words?
- Could I explain spherical excess to someone else?
- If the Earth were flat, how would I explain why surveyors consistently measure excess in large triangles?

If you can explain how surveyors measure excess in long-distance triangles while maintaining that it is not caused by curvature, your skepticism is informed.

If you reject curvature without explaining the measured excess, your conclusion may be guided by confirmation bias rather than evidence.

If these explanations make sense to you, try explaining the concept to someone else. Teaching it is one of the best ways to confirm your understanding.

# Chapter 8

## Navigation: How Long-Distance Travel Works on a Sphere

In the previous chapter, surveying revealed spherical geometry through measured angles that accumulate spherical excess over large areas. Navigation provides a second independent measurement system that tests the same geometric conclusion revealed by surveying. Navigation reaches the same conclusion in a different way—not by measuring angles across land, but by solving a practical problem of distance and direction:

How do we reliably travel thousands of miles and arrive exactly where navigation predicts?

Like surveying, navigation is not theoretical. It either works or it fails. Over long distances, even small geometric errors grow large. If the wrong model of Earth were used, ships and aircraft would miss their destinations by measurable and accumulating margins.

They do not.

### What Navigation Measures

Navigation depends on measured quantities, not appearances. It requires:

• Direction (bearings and headings)
• Distance traveled
• Position relative to Earth's surface
• Time, when using celestial methods

Over short distances, flat approximations are often adequate. Over thousands of miles, geometric differences become measurable.

This makes navigation a direct measurements-based test of Earth's shape.

It therefore provides another way to test the premise introduced in Chapter 5: that independent measurement systems agree when Earth is modeled as a sphere.

## Shortest Paths: Straight Lines or Great Circles?

On a flat surface, the shortest path between two points is a straight line. On a flat map, that path appears straight.

However, long-distance flight paths on flat maps often appear curved. This can make it appear as if aircraft are not taking the shortest route. That appearance is misleading: the curvature reflects distortion in the map projection, not the aircraft's actual path, which follows a great-circle route.

On a sphere, the shortest path between two points is a great-circle path—a circle whose center coincides with the center of the sphere. When such a path is represented on a flat map, it must appear curved.

If Earth is flat, long-distance routes should follow straight-line geometry across a flat plane.
If Earth is spherical, long-distance routes should follow great-circle geometry across a sphere.

This is not a matter of visual interpretation. It is a measurable difference in distance, time, and direction.

In practice, ships and aircraft consistently follow great-circle routes because those routes minimize distance and match real-world navigation—demonstrating that spherical geometry, not flat geometry, governs long-distance travel.

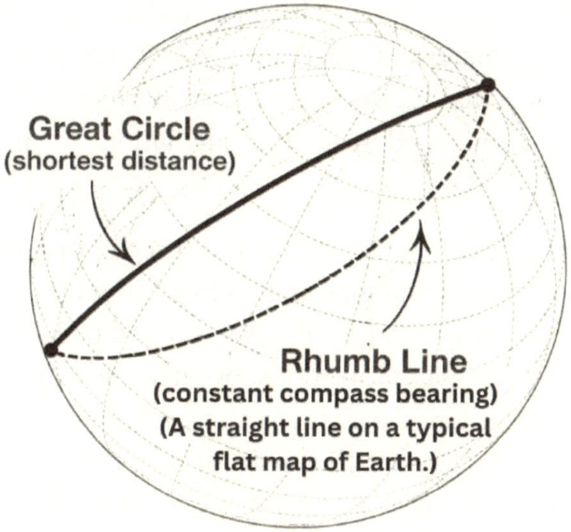


## Demonstration: Route Comparison

Think about a long transoceanic flight—for example, a route from New York to Tokyo. The flight follows a Great Circle route that, on most flat projection maps of Earth, is shown to curve northward over Alaska and the North Pacific. This curved route looks longer than needed on a typical flat map, but when plotted on a globe, it looks almost straight from city to city.

If pilots were to instead use a flat projection map and follow a straight-line path on the map, in order to maintain a constant compass bearing—known as a rhumb line—without accounting for spherical geometry, two measurable problems would occur:

The total distance and time traveled along that constant compass bearing would be much longer than necessary.

Maintaining a constant compass bearing would gradually carry the aircraft off the planned route, causing it to miss the intended arrival coordinates.

In practice, flight management systems do not use planar (flat) geometry. They calculate great-circle routes based on a spherical (more precisely, oblate spheroid) Earth model. The predicted distances match actual fuel usage and arrival times, and thousands of aircraft arrive when and where the calculations predict they will every day.

Modern air travel daily demonstrates the agreement between spherical geometry calculations and reality. Public flight maps routinely display long-distance routes as curved paths rather than straight lines on flat map projections, because the routes being calculated are great-circle paths on a curved Earth. Although these routes appear curved on two-dimensional maps, they represent the shortest practical distances over Earth's curved surface. The consistent success of these calculations in guiding aircraft accurately and efficiently all around the world provides measurable, repeated confirmation of the spherical geometry used in modern navigation.

## Continuous Bearing Change

Another measurable effect distinguishes flat from spherical geometry.

On a sphere, a great-circle path does not maintain a constant compass bearing except along the equator or along meridians. Instead, the heading must continuously adjust.

Modern autopilot systems account for this automatically. Navigational software constantly changes the aircraft's bearing to remain on the great-circle route.

If Earth were flat, no such continuous correction would be required for the shortest path. A constant heading would suffice.

But in real-world navigation over long distances, heading adjustments are necessary—and calculated precisely using spherical geometry.

## Celestial Navigation Before Modern Technology

Long before satellites, sailors navigated using angular measurements of the sky. (A homework video explains this well.)

In the Northern Hemisphere, the altitude of Polaris—the North Star, which appears nearly fixed in the sky and marks true north—corresponds directly to the observer's latitude. If Polaris appears 40 degrees above the horizon, the observer is approximately at 40° north latitude.

This relationship only holds if:
• Earth is spherical
• lines of latitude, such as the equator, are circular
• the observer's vertical direction—toward Earth's center of mass—changes with location

On a flat Earth, Polaris could not simultaneously appear at different elevations to observers separated north-to-south in a way that scales linearly with north–south distance.

Longitude was determined using time. Because Earth rotates 360 degrees in 24 hours, each hour corresponds to 15 degrees of rotation. By comparing local noon to the time at a reference location, navigators calculated their east-west position.

These ancient calculations depended on spherical geometry and Earth's rotation. They were not approximate guesses—they were repeatable measurements that allowed ships to cross oceans and return.

Today, ships use electronic navigation systems, including satellite-based positioning, that rely on the same underlying geometry.

If spherical geometry were wrong, ships that use it for navigation would not repeatedly reach their destinations.
They do.

On a flat Earth, long-distance air and sea routes would appear as straight lines on flat maps.
They don't.

## Why Flat Geometry Fails

If Earth were flat:

• great-circle routes would not consistently minimize travel distance

• long-distance routes would not require continuous bearing adjustments

• the altitude of Polaris would not correspond directly to latitude

• navigational tables based on spherical geometry would not agree with real voyages. Instead, navigational calculations agree across centuries of practice, across oceans, and across independent methods.

Non-spherical models have never produced a complete, quantitative navigational system that matches global travel without quietly reintroducing curvature under a different name.

Navigation is unforgiving.
It cannot survive on appearances or intuition alone.
It works specifically because the underlying geometry matches reality—spherical.

## What Navigation Demonstrates

Navigation reveals spherical geometry through great-circle routes.
It does not rely on videos or photographs.
It does not rely on trusting institutions or authorities.
The measurable quantities must agree if travel is to succeed.
And they do agree—only when the Earth is modeled as a sphere.

That makes navigation another system where large-scale geometry quietly reveals itself.
It works because the spherical geometry it uses matches the measured shape of Earth.

This consistency in navigation sets the stage for another independent measurement system: radio communication.

In radio communication, curvature appears not in angles or routes, but in the measurable limits of signal transmission itself.

# Homework

## Understand Navigation According to Both Models

### Watch: Spherical Model

"Great Circles and Rhumb Lines – Types of Routes" by: Aviation Theory youtube.com/watch?v=3BF_ZKfJiso

"Why Don't Ships Go Straight? | Great Circles" by: Casual Navigation youtube.com/watch?v=u8A3yAbcMcM

"Great Circle Route Copy" by: Tom Ferguson youtube.com/watch?v=KzYsb3Bys9Y

"The GENIUS of Inertial Navigation Systems Explained" by: FlyByMax youtube.com/watch?v=Pq_PDaYclAw

While watching "The GENIUS of Inertial Navigation Systems Explained," ask yourself why anyone would design such a sophisticated navigation system if the Earth were not a sphere.

"Polaris and Latitude" by: Russ Johansen youtube.com/watch?v=gdRU2vyBWqo

You may also want to search YouTube for:

- "How great-circle navigation works"
- "Why airplane routes curve on maps"

Watch as many videos as needed to ensure you understand the spherical-Earth explanations.

Then, deliberately seek out flat-Earth explanations of long-distance navigation, including how flight paths, great-circle routes, and inertial navigation systems are interpreted within a non-spherical model. Compare those explanations directly with the routes aircraft actually follow and the measurements navigation systems rely on.

## Investigate

As you watch, try to answer these questions:

- What route does the spherical model predict between two cities?
- What distance should that route be?
- Do aircraft actually follow that predicted path?
- Can you restate the explanation fairly in your own words?

**Treat this like a small research project.**

Search the Internet for images of "air traffic flight maps." Scroll through the images and ask yourself why the lines representing the flight routes appear curved.

If you have a globe, take a piece of paper and wrap it around the globe so that one edge forms a straight line between New York City and Tokyo. Then imagine flattening the globe into a map while trying to preserve that same path.

Would the line appear curved or straight on the flat map?

## Reflect

Ask yourself:

- Can I clearly explain the difference between great-circle routes and rhumb lines?
- Why does the shortest path between two points on a sphere appear curved on most flat maps?
- Could I explain this concept to someone else?

If you can explain the difference between great-circle routes and rhumb lines clearly and still reject the spherical-Earth model, your evaluation is informed.

If you reject the model without understanding that difference, your decision may be guided more by confirmation bias than by evidence.

If the explanations in these videos make sense to you, try explaining the concepts to someone else in your own words. Teaching an idea to another person is one of the best ways to verify that you truly understand it.

# Chapter 9

## Radio Communication: Measuring the Radio Horizon on a Sphere

Surveying revealed curvature through angle sums. Navigation revealed curvature through route geometry. Radio communication reveals curvature in a different way: through measurable limits on how far signals travel.

Radio engineers do not ask what the Earth looks like. They ask a practical question that everyone who listens to the radio eventually discovers the answer to:

At what distance will this signal stop being received?

The answer depends on geometry.

Radio systems must function reliably. Engineers must be able to predict where reception will be strong, where it will weaken, and where it will disappear entirely. If those predictions were based on the wrong geometric model of Earth, radio networks would fail to operate predictably.

Instead, radio communication works with remarkable reliability across the world.

The reason is simple: the calculations used to design radio systems match the Earth's spherical geometry.

## What Radio Engineering Actually Tests

Radio transmission involves electromagnetic waves traveling from a transmitting antenna to a receiving antenna. Engineers must predict:

- how far the signal will travel
- where reception will cease
- how antenna height affects coverage

To make those predictions, they measure:

- antenna height
- receiver height
- frequency
- terrain
- Earth's curvature

Notice what is absent: visual appearances.

Radio engineers do not rely on how the landscape looks. They rely on measurements and spherical geometry.

The calculations used in radio engineering must match reality closely enough for communication networks to function. Broadcast stations, emergency communication systems, aircraft radios, and marine radios all depend on those predictions.

If the geometric assumptions were wrong, engineers would not be able to design reliable communication networks.

## The Radio Horizon

The central geometric concept is the radio horizon.

If Earth were flat, there would be no geometric horizon. Signal range would be limited only by power, interference, or terrain, and increasing transmitter power would indefinitely extend that range.

But that is not what engineers measure.

Instead, radio signals are constrained by a geometric limit: the horizon blocks their path. Range is therefore limited not only by power, but also by the Earth's curvature.

Consequently, signal range scales predictably with antenna height—exactly as spherical geometry predicts.

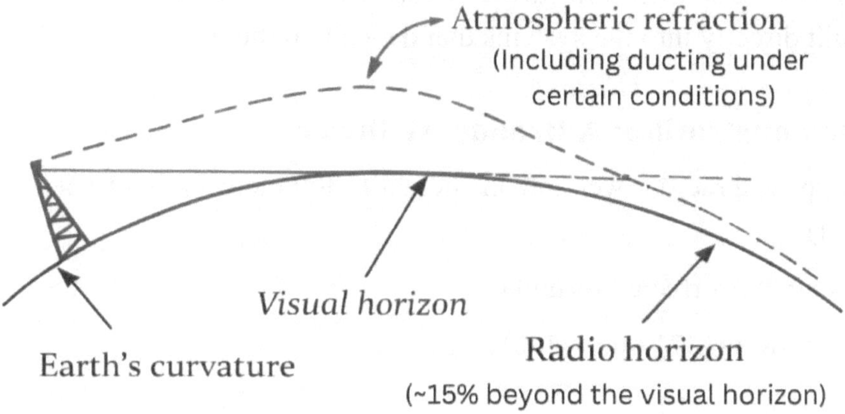


## The Geometry

For an antenna of height h above a spherical Earth of radius R, the geometric horizon distance d is:

$d \approx \sqrt{(2Rh)}$

Because radio waves bend slightly downward in the atmosphere, engineers use an effective Earth radius correction, commonly approximated as:

$R_{e}ff \approx 4/3\ R$

This modifies the horizon formula to:

$d \approx \sqrt{(8Rh / 3)}$

This adjustment is not arbitrary. It reflects measured atmospheric refraction that slightly bends radio waves downward as they travel through the atmosphere.

Engineers apply this correction consistently in communication system design.

Claims that these observations contradict a spherical Earth often overlook that these effects are already measured, modeled, and built directly into the systems that depend on them.

## Demonstration: A Broadcast Tower

Suppose a radio tower antenna height is 300 meters (about 1,000 feet).

Using the corrected formula:

$$d \approx \sqrt{((8 \times 6{,}371{,}000 \times 300) / 3)}$$

$$d \approx \sqrt{(5.1 \times 10^9)}$$

$$d \approx 71{,}000 \text{ meters}$$

$$d \approx 71 \text{ km} \approx 44 \text{ miles}$$

This predicts that a low car antenna should lose reception around 40–45 miles away under normal conditions.

That is exactly what drivers experience.

As distance approaches that range:

- signal weakens
- static appears
- reception disappears

Turning up the volume does not restore the signal. Increasing transmitter power does not extend range indefinitely.

Only raising antenna height extends the measurable horizon.

This behavior matches spherical geometry—not flat geometry.

## Two-Antenna Calculation

Now include both antennas:

Tower height: 300 meters
Car antenna height: 2 meters

Each antenna has its own geometric horizon.

Tower horizon:

$d_{tower} \approx 71$ km

Car horizon:

$d_{car} \approx 6$ km

Total line-of-sight range:

71 + 6 = 77 km

≈ 48 miles

Again, this matches real-world experience with remarkable consistency.

This is not a visual effect. It is a geometric constraint caused by Earth's spherical shape.

## Why Power Cannot Defeat Curvature

If Earth were flat, signal strength would decline gradually with distance but could be compensated by increasing power. There would be no sharp geometric cutoff determined by antenna height.

But once a receiver drops below the radio horizon, the signal path is physically blocked by Earth's curvature.

No amount of additional power restores it.

The only solutions are:

- raising the transmitting antenna
- raising the receiving antenna
- installing relay towers

All three solutions are based on the same geometric assumption: Earth's surface curves.

Relay towers, microwave links, and communication satellites all exist because engineers must account for that curvature when designing communication systems.

## Exceptional Conditions

Under certain atmospheric conditions, radio signals can travel farther than normal—a phenomenon known as ducting.

Engineers measure and account for atmospheric refraction, including ducting, because it modifies signal paths without removing the underlying geometric limit.

Ducting does not eliminate the geometric horizon; it slightly alters the effective curvature experienced by the signal. Even so, the behavior of radio transmission, under typical conditions, still depends on spherical geometry.

## What Radio Communication Demonstrates

Radio engineering provides a clean geometric test.

Measurements suggest that:

- signal range scales with antenna height
- the scaling matches spherical formulas
- refraction modifies but does not eliminate curvature
- increasing power cannot overcome the horizon
- independent measurements align with a specific Earth radius

This pattern is not accidental. It reflects a consistent underlying geometry—spherical.

Radio communication therefore provides another independent system in which measurements, predictions, and real-world performance all agree. This agreement does not arise from assumption, but from repeated measurements.

In the next chapter, we turn to perhaps the most precise and globally distributed measurement system ever built: GPS. There, curvature is not inferred from angles or horizons, but calculated directly through distance equations in three-dimensional space.

# Homework

## Understand Radio Horizons According to Both Models

### Watch: Spherical Model

"Radio Waves" by: TheOnlineEngineer
youtube.com/watch?v=sRX2EY5Ubto

"Radio Horizon" by: Ku Siti Syahidah Ku Mohd Noh
youtube.com/watch?v=mUSGFfVIK5o

"Understanding VHF Propagation" by: Rohde & Schwarz
youtube.com/watch?v=DTXJ3bS2UcE&t=452s

"Space Wave Propagation (Basics, Limitations, Radio Horizon Extension & Applications) Explained" by: Engineering Funda
youtube.com/watch?v=eNF7aIt4B6g

You may also want to search YouTube for:

- "Radio horizon explained"
- "How antenna height affects radio range"

Watch as many videos as needed to ensure you understand the spherical-Earth explanations.

Then, deliberately seek out flat-Earth explanations of radio propagation and signal range. Compare those explanations directly with the distances engineers actually measure and predict using antenna height and radio horizon calculations.

### Investigate

As you watch, try to answer these questions:

• What distance does the spherical formula predict for a given antenna height?
• Do engineers report signal ranges that match that prediction?
• How does increasing antenna height change the distance a signal can travel?
• Does this help explain why you cannot listen to your local radio station while far out of town?

Pay attention to the relationship between antenna height and signal range, and how engineers use spherical geometry to predict where reception will stop.

## Reflect

Ask yourself:

• Why do radio stations build extremely tall broadcast antennas?
• What problem are those towers designed to overcome?
• Could I explain the relationship between antenna height and signal distance to someone else?

If you can explain why radio stations spend millions of dollars building extremely tall broadcast antennas because engineers believe Earth's curvature requires them—while still believing there is no curvature—then your skepticism is informed.

If you reject the spherical model without understanding how radio horizon and antenna height determine signal range, your conclusion may be guided more by confirmation bias than by evidence.

If the explanations in these videos make sense to you, try explaining the concepts to someone else in your own words. Teaching the idea to another person is one of the best ways to confirm that you truly understand it.

# Chapter 10

## GPS: Measuring Position on a Spherical Earth

Surveying revealed curvature through angle sums.
Navigation revealed curvature through route geometry.
Radio communication revealed curvature through measurable horizons.

Global Positioning Systems—GPS—reveal curvature in a more direct and powerful way: by calculating distances in three-dimensional space and placing our phones and cars precisely on Earth's surface, within a few meters, anywhere on the planet.

GPS does not rely on appearances.
They rely on measured time, calculated distance, and three-dimensional geometry.
And they work because Earth is treated as a sphere with a measurable radius.

### What GPS Actually Measures

In GPS, trilateration uses measured distances.

GPS determines position by measuring the distance from a receiver, such as a car or phone, to at least four satellites. Let's use your phone as an example.

Each satellite broadcasts its precise orbital position and the precise time each signal is sent. Your phone uses a mathematical calculation to determine how long each signal takes to arrive. By comparing the arrival times of signals from multiple satellites, the receiver calculates its distance from each satellite.

Because radio waves travel at the speed of light, the calculation looks like this:

distance = (signal travel time) x (speed of light)

where:

speed of light ≈ 299,792,458 m/s (about 300 m per microsecond)

Note that even a timing error of one microsecond would produce a location error of nearly 300 meters. That level of sensitivity means GPS is not forgiving. Its geometry must be correct.

The system therefore relies on extremely precise clocks and carefully calculated satellite orbital positions. If the geometry used in those calculations were wrong, the errors would accumulate quickly and visibly.

## Trilateration in Three Dimensions

Your phone's distance is measured from each of at least four satellites, and every measurement is considered the radial distance of a sphere that is centered on each satellite. In other words, the measured distance to each satellite defines the radius of a sphere around that satellite, and you are located somewhere on the surface of that sphere. When distances to multiple satellites are calculated, each defines the radius of a sphere around that satellite, then the exact point where all of the satellite spheres overlap is your phone's position.

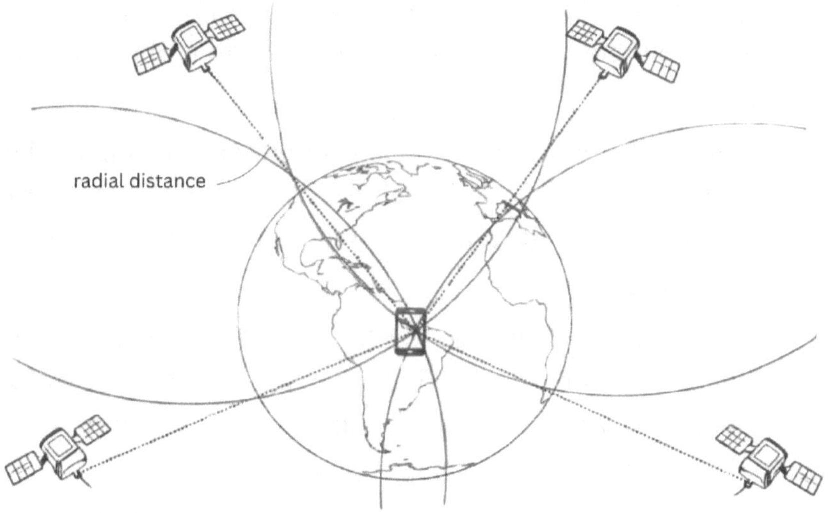


I want to clearly explain how it works, so I'm going to describe it again using other words.

The distance from your phone to each satellite is calculated as the radius of a sphere centered on that satellite. The single position where at least four satellite spheres overlap is mathematically calculated to be your position in space—specifically, on Earth.

• One satellite → you are somewhere on a sphere.
• Two satellites → you are somewhere on the circle created by two overlapping spheres.
• Three satellites → you are at one of two overlap points.
• Four satellites → clock error is corrected and your exact three-dimensional location in space is determined.

This process is called trilateration.

Using at least four distances—and factoring in orbital mechanics, the Earth's center of mass, and the Earth's gravitational field—your phone is programmed to solve equations that look like this:

$\sqrt{[(x - x_i)^2 + (y - y_i)^2 + (z - z_i)^2]} = c\ (t_i - t_r)$

Such equations—which your phone is solving every time it determines your position—assume that the Earth is approximately spherical.

If the planet had a radically different geometry (for example, flat), the predicted orbital paths and satellite positions relative to Earth's surface would not match reality—and the equation would not determine your position.

GPS equations are solved in an Earth-centered coordinate system, meaning positions are calculated relative to a central point and a consistent radius. In practice, this requires modeling Earth as a spheroid with a measurable center of mass.

GPS uses a model called "World Geodetic System 1984," which treats Earth as an oblate spheroid (a slightly flattened sphere). Satellite orbits, ground stations, and coordinate calculations are all based on this same model.

This is another verification of mathematical agreement across different systems: if spherical geometry were wrong, the equations would not reconcile globally—the satellite distance spheres would not intersect consistently across continents and oceans to produce stable positions.

But they do. Your phone can determine your location within a meter or two anywhere on Earth.

The details can be complex, but the key point is simple: these GPS calculations only work because Earth behaves as a sphere.

Whenever you use GPS, your position is being calculated from measured distances to multiple satellites and expressed as latitude, longitude, and altitude within this global coordinate system.

The companies that design phones, navigation systems, and satellite networks use this geometry not to defend a theory, but because it allows their technology to function reliably. That reliability is not a matter of opinion—it is a continuous, global test of whether the mathematics matches reality.

This is an important point that extends beyond GPS.

When these systems depend on precise measurements and produce consistent, repeatable results, they become a test of the spherical assumptions built into them. The more often they are seen working, the more they verify those assumptions.

Claims about how such systems work must therefore be evaluated not by how persuasive they sound, but by whether their underlying premises match the measurements and mathematics that make the systems function as designed.

Large aerospace companies such as Northrop Grumman, Boeing, Lockheed Martin, and Airbus Defence and Space invest billions of dollars designing and manufacturing satellites. Private companies such as SpaceX have launched thousands of satellites through systems like Starlink. These systems are not theoretical—they are operational.

It is reasonable to ask: why would such systems be built and maintained at this scale if the underlying geometry did not correspond to reality?

Instruments such as the Hubble Space Telescope and the James Webb Space Telescope continuously collect data and images that are publicly available. These observations are part of a broader network of measurements that align with orbital mechanics and spherical geometry.

Taken together, these systems do not merely suggest a spherical Earth—they depend on it.

## Satellite Motion Requires Curvature

GPS satellites orbit about 20,200 kilometers (≈12,550 miles) above Earth's surface and complete an orbit roughly every 12 hours.

Their motion is predictable because:

- they move around a central body—the Earth
- at a known radius
- under gravitational attraction

The system depends on precise orbital mechanics. The satellites' positions must be known to within meters. Those positions are calculated using Earth's radius, gravitational field, and three-dimensional geometry.

Satellite motion is calculated relative to Earth's center of mass and gravitational field, and those mathematical calculations produce the stable, predictable satellite trajectories required for global timing and positioning. If the Earth were not approximately spherical, there would be no single Earth radius by which satellite altitude could be defined consistently. The orbital calculations and timing models that GPS relies on would no longer reconcile with the measured positions of satellites, and the positioning system built on those calculations would fail. It does not.

## Relativity: Precision Beyond Intuition

GPS is so precise that two relativistic corrections are required:

Special relativity (satellites move fast → clocks run slower):
−7 microseconds per day

General relativity (satellites are higher in a weaker gravitational field → clocks run faster):
+45 microseconds per day

Net effect:
+38 microseconds per day

If this correction were ignored, GPS position errors would accumulate at roughly 10 kilometers per day.

These corrections depend explicitly on:

- Earth's mass
- Earth's radius
- satellite altitude

They are not aesthetic adjustments. They are measured necessities.

## Why Non-Spherical Earth Models Fail

Non-spherical Earth models cannot produce a single global coordinate system in which the following remain true:

- satellites orbit predictably
- signal timing matches distance calculations
- position equations intersect consistently
- relativistic corrections scale correctly with altitude

If Earth were not a sphere:

- trilateration equations would not reconcile across hemispheres
- distance errors would vary unpredictably by location
- satellite altitude would have no reference radius
- GPS accuracy would degrade rapidly and measurably

Instead, GPS works everywhere—equator, poles, oceans, continents—with meter-level accuracy.

In short, your phone can determine your location anywhere on Earth because the underlying geometry matches reality—spherical.

## A Direct Observation: Starlink Satellites

If you would like to test the accuracy of orbital prediction in action, you can look up and confirm that predicted trajectories match what you see by watching Starlink satellites pass overhead.

Publicly accessible websites such as **findstarlink.com** allow you to enter your location and receive a precise time window—often within a few minutes—during which satellites will pass across your sky, including predicted direction and elevation angles.

Publicly announced orbital predictions are not broadcast to persuade—they make it possible for anyone to go outside on clear evenings or mornings, look up, and repeatedly confirm those predictions for themselves.

These predictions are possible because satellite positions are calculated using spherical Earth geometry and orbital mechanics.

Claims that satellites are fictitious must explain not only the claims themselves, but also the repeatable success of public orbital predictions.

If Earth were not a sphere with measurable radius and gravitational center, orbital paths would not be predictable in this way, and such timing accuracy would not exist.

The predictions are accurate because they correspond to reality.

What you observe directly in the sky aligns with what multiple independent measurement systems consistently indicate.

## Independent Confirmation

GPS does not depend on:

- surveying triangles
- navigation charts
- radio-horizon calculations

It uses different instruments, different physics, and different mathematics than those systems.
Yet its measurements align with the same spherical geometry and therefore lead to the same conclusion:

**Earth behaves as a sphere with a measurable radius.**

This consistent agreement across all of these systems should remind us of our primary goal when evaluating arguments.

As humans, we are not neutral observers. We are inclined to accept claims that fit what we already believe—and to repeat those claims without testing them carefully.

For that reason, we need to remember that our goal is to determine not which argument sounds most persuasive, but which has premises that actually match the measurements that make so many systems work.

## What GPS Ultimately Demonstrates

What you can verify directly in the sky—that Earth behaves as a sphere—is not an isolated observation.

Surveyors measure curvature in angles.
Navigators measure curvature in routes.
Radio engineers measure curvature in signal horizons.

GPS directly measures curvature through three-dimensional distance equations and orbital motion.

GPS does not ask you to trust what you see.
It does not ask you to accept photographs.
It does not ask you to trust institutions.

GPS is not just one measurement. It is a network of continuously cross-checked measurements:

- Satellite tracking stations
- Orbital predictions
- Atomic clock synchronization
- Distance trilateration
- Ground verification

Those different technologies and mathematical models all agree using spherical geometry.

GPS solves equations based on measured time and known physical constants—and those equations reconcile globally only when Earth is modeled as a sphere.

The equations that determine position assume that Earth's surface lies at a nearly constant radial distance from Earth's center of mass. That condition is exactly what defines a sphere or spheroid.

If Earth were not a sphere, GPS would fail globally and immediately.

It does not.

GPS does not merely confirm that Earth is a sphere as a technical detail. It represents something deeper: a global, constantly operating measurement system that works quietly and consistently because Earth is spherical, whether we believe that or not.

Every time a position is calculated, every time a phone determines its location using satellites, every time distance equations reconcile across continents and oceans, the same spherical geometry holds.

Consequently, we should no longer be asking whether one observation can tell us the shape of the Earth. We should be asking what conclusion most logically follows from the fact that so many independent measurement systems—developed for entirely different purposes—all agree on spherical geometry.

When independent systems measuring different aspects of reality agree mathematically, the simplest explanation is that the geometry they share describes the world correctly.

Different measurements.
Same world.
Same spherical geometry.

In addition, across these systems, claims that appear to challenge the model are consistently explainable within well-established scientific frameworks rather than requiring a different geometry of the Earth. This explanatory power is itself evidence in support of the spherical model.

In the final chapter, we will step back and think about what this consistent agreement across independent lines of evidence means—not just for the shape of the Earth, but for how we decide what to trust when appearances and measurements disagree.

# Homework

## Understand GPS and Satellites According to Both Models

### Watch: Spherical Model

"GPS: How It Finds You" by: On The Shoulders Of Science youtube.com/watch?v=qPTIi7Ds15M

"GPS, How Does It Work? | ICT #12" by: Sabin Civil Engineering youtube.com/watch?v=8eTlI19_57g

"How Does GPS Work?" by: sciBRIGHT youtube.com/watch?v=FU_pY2sTwTA

"SpaceX Unveils Insane New Product" by: Farzad youtube.com/watch?v=sly91aVXWQ8

"Starlink Mobile Is Now the Largest Cell Network On Earth" by: JCristina youtube.com/watch?v=HAogxFe-IGI&t=88s

What explanation best accounts for videos describing satellite-based systems like Starlink Mobile?

Take the time to understand the spherical-Earth explanations provided in the videos below.

You may also want to search YouTube for:

- "How GPS trilateration works"
- "Why GPS requires relativity corrections"
- "How GPS works"

Watch as many videos as needed to ensure you understand the spherical-Earth-model explanations.

Then, deliberately seek out flat-Earth explanations of GPS, satellites, and orbital systems. Compare those explanations directly with how GPS actually calculates position, how satellites are tracked and predicted, and how these systems function in everyday use.

A related topic to investigate involves long-distance sailing routes around Antarctica. Watch this video:
"Antarctica causes so many problems for Flat Earth" by: Dave McKeegan youtube.com/watch?v=TmHVVZBmHTM

Then watch this short video presenting an alternative claim:
"Antarctic Sail Boat Race : From Cape Town to Brazil"
youtube.com/shorts/3aZtdCiv9UE

As you compare these perspectives, ask:
• What routes do long-distance sailing races actually follow around Antarctica?
• Do multiple independent races and voyages circumnavigate the continent?
• Which explanation better accounts for recorded navigation paths and voyage data?

In the second video, it is claimed that "NASA runs GPS."
Investigate this claim:
• Which organizations actually operate satellite navigation systems such as GPS?
• Are there multiple independent global navigation systems?
• Can their accuracy be verified directly through everyday use?

As you evaluate these claims, focus on which explanation best accounts for the available evidence. And remember: The goal is not to accept a claim, but to determine which explanation best matches measurable reality.

This is not about winning an argument or proving someone wrong. It is about discovering what is true. The goal is not to defend a position, but to test it—to follow the evidence wherever it leads and to be willing to change our minds if the measurements require it.

## Investigate

As you watch, try to answer these questions:

- How does your GPS receiver determine its distance from satellites?
- Why are at least four satellites needed to determine your position?
- Why are GPS positions first calculated in a three-dimensional Earth-centered coordinate system before being converted into latitude, longitude, and altitude?
- Could these spherical-geometry equations produce a consistent solution on a non-spherical Earth model?
- Could you explain GPS trilateration simply to someone else?

Pay attention to how GPS relies on measured time, calculated distance, and spherical geometry to determine position.

Visit the website **findstarlink.com** and enter your location. The site will show you dates and times when Starlink satellites will pass overhead. Choose a time labeled "BRIGHT" for best visibility. At the recommended time, go outside and look in the direction indicated on the website. You may see a line of lights moving steadily across the sky. Using binoculars can make them easier to see.

I have personally observed many lines of satellites passing overhead at exactly the predicted times.

See it for yourself and ask: how could such precise predictions be possible if satellites are not actually in orbit?

Also, visit the websites of the following organizations:

• Northrop Grumman-- northropgrumman.com
• Boeing-- boeing.com
• Lockheed Martin-- lockheedmartin.com
• Airbus Defence and Space-- airbus.com
• SpaceX and Starlink-- starlink.com

As you explore, ask and consider:

• What do these companies build?
• How many people do they employ?
• How many satellites have been launched?
• What systems depend on those satellites?
• Why would companies like SpaceX and its subsidiary Starlink tell the world they are building and launching satellites into orbit if they were not actually doing it?

Which story best explains the scale and coordination of these multibillion-dollar companies:
• real satellite engineering and deployment
• or a global effort to simulate it

Which explanation better accounts for the existence and operation of such massive companies?

The goal is not yet to decide on a conclusion, but to understand which explanation best accounts for the measurements.

## Reflect

If you can explain how GPS works based on the spherical Earth model, yet still provide a better explanation of how our phones can be precisely located on a non-spherical Earth, then your understanding is well tested.

If you reject the spherical model without understanding how GPS locates your phone and car, your conclusion may be guided more by confirmation bias than by evidence.

If the explanations in these videos make sense to you, try explaining the concepts to someone else in your own words. Teaching the idea to another person is one of the best ways to confirm that you truly understand it.

# Chapter 11

## Conclusion

In Chapter 5 we stated a measurements-based argument about the shape of the Earth and began testing its premises.

The previous chapters have examined several independent measurement systems that operate across the Earth. In the last chapter, we looked at one of the most precise of them: GPS, a system so exact that nanoseconds matter. GPS does not rely on perspective, photographs, videos, or appeals to authority. It works through measured time, calculated distance, orbital mechanics, and three-dimensional geometry. And it works everywhere—quietly, consistently, and regardless of our beliefs.

With measurements this precise, we should no longer be focused on whether a distant city appears visible across water or whether the horizon looks straight. Those appearances have been explained and can be predicted.

Instead, after seeing so many measurements from different systems agree, one question becomes unavoidable:

What should we conclude when independent measurement systems that were developed for entirely different purposes all mathematically agree on the same spherical geometry?

To answer that, we must return to where we began—not to start over, but to understand it more clearly.

From where we stand, Earth appears flat.

That appearance is not foolish. It is not naïve. It is not the result of ignorance. It is exactly what we should expect from beings who live at human scale, whose minds create perspective-based imaginations that help us navigate fields, roads, buildings, and shorelines—not to comprehend planetary geometry.

When we walk outside, the ground feels level.
When we look at the horizon, it appears straight.
When we stand still, we do not feel motion.

Nothing in our immediate experience requires us to imagine curvature or rotation.

That is why non-spherical claims feel powerful: they begin with something real.

But in Chapter 2, we saw that not all appearances reflect underlying reality. Some appearances are illusions—false imaginations produced by the mind from real sensory input. Perspective compression, contextual cues, geometric patterns, and even the bending of light can all lead our minds to construct a picture that does not match the true structure of the world.

The appearance of flatness is one such case. It is not a deception in the sense of being fabricated—it is a natural consequence of how human perception works at large scale.

But this book has not asked whether Earth appears flat to us. It has asked something more challenging:

What happens when we test appearances—including those shaped by illusion—against measurements?

Appearances and measurements do not always agree.
To understand what that means, we must look beyond any single observation and examine what measurements reveal across many independent systems.

When appearances and measurements conflict, we face a choice. We can trust what feels obvious—or we can trust what has been carefully measured and repeatedly confirmed.
We can either treat appearances as final—even when we know that appearances can include illusions—or allow independently verified measurements, tested across multiple independent fields, to correct what our senses cannot reliably determine at planetary scale.

The measurement systems that guide aircraft, ships, radio communication, satellites, and global navigation networks do not rely on appearances.

Machines do not care what Earth looks like to us. They do not hold beliefs, and they do not defend conclusions. They simply perform calculations based on measured quantities such as time, distance, angle, and signal travel time. The companies building those machines are not trying to persuade us of a theory about Earth's shape; they are aiming to build products that work.

As we asked in Chapter 5: "how many independent measurements, repeated across how many independent systems, must agree before it becomes unreasonable to disbelieve what they consistently indicate—that the shape of the Earth is spherical?"

Here is what we have examined:

• Surveyors measure angles across large regions and discover spherical excess that grows precisely with the area of the triangle—exactly as spherical geometry predicts.
• Navigators calculate long-distance routes and find that great-circle geometry, not flat-plane geometry, matches real travel over a sphere with Earth's radius.
• Radio engineers measure horizon limits and find that signal range scales predictably with antenna height—exactly as spherical geometry predicts.
• GPS solves three-dimensional distance equations using orbiting

satellites and finds that those equations reconcile globally only when Earth is treated as a sphere with a measurable radius.

What we have examined in this book is only a small portion of the available evidence—evidence that extends across many additional fields, including astronomy, satellite imaging, geodesy, physics, and engineering.

In the previous chapters, you have been prepared to make a fully informed decision. You have examined how multiple independent measurement systems function in practice, compared competing explanations, and seen which model consistently matches what is measured.

In different fields and systems, the same spherical geometric shape emerges consistently—not as illustration, but as direct empirical support that establishes the premises of the argument as true.

Because the argument is valid and the premises have been established, the conclusion follows.

Earlier in this book, you were encouraged not to decide on these conclusions without first testing their premises for yourself. The purpose of the homework was not to persuade you, but to give you the opportunity to examine both models against the same measurements. It required being willing to say to yourself, "Maybe this argument for a spherical Earth is true—and maybe my non-spherical beliefs are wrong."

If you have done that—carefully and honestly—then what we are seeing here is not a claim being asserted, but a conclusion tested against measurements.

Different measurements.
Independent methods.
The same geometric result.

And in every case, that geometry is spherical.

That geometry is not merely aesthetic.

It is mathematical.
It is testable.
It is repeatable.
It is global.

## Returning to the Argument from Chapter 5

In Chapter 5 we stated a measurements-based modus tollens argument about the shape of the Earth.
Its reasoning was straightforward.

**Premise 1 — Conditional**
If the Earth were not spherical (that is, if it were flat or some other non-spherical shape), then independent measurement systems that rely on spherical geometry would fail to agree mathematically with one another.

**Premise 2 — Negation of the Consequent**
But these independent measurement systems—surveying, navigation, radio propagation, and global positioning—developed for different purposes do agree mathematically, consistently, predictably, and precisely when Earth is modeled as a sphere with a specific measurable radius.

Because the logic of the argument is valid and its premises are supported by independent measurements, the argument is not only valid but sound.

And when an argument is sound, its conclusion cannot be false.

**Conclusion — Modus Tollens**

***Therefore, the Earth cannot be flat or any other non-spherical shape.***

This argument does more than answer the question about the shape of the Earth.

It demonstrates a method for evaluating any claim.

When an argument is valid and its premises are carefully tested and shown to match reality, its conclusion is not optional.

And when a claim depends on premises that have not been tested—or that contradict established measurements—it should not be accepted, no matter how persuasive it feels.

We are free to learn when we do not allow our emotional attachment to beliefs to drive us to defend them against contrary evidence, but instead use that evidence to test them.

## When Belief Becomes a One-Way Door

There is a psychological moment in many belief systems—a moment when a conclusion stops being a hypothesis and becomes a commitment, when it becomes difficult to imagine what new evidence could change it.

At that point, evaluating a claim shifts from:

"Is this true?"

to

"This must be true—so how do I explain everything else?"

When someone becomes convinced that the Earth cannot be a sphere, based primarily on what it looks like from ground level, something subtle can happen. Every contrary measurement must then be reinterpreted or rejected, or attributed to deception—not because the measurements are unclear, but because the belief cannot accommodate them. Entire systems—surveying, navigation, radio engineering, global positioning—must be treated as coordinated error or conspiracy.

It insulates us from testable evidence.

Once the “Is this true?” door closes, it becomes increasingly difficult to evaluate conflicting evidence honestly. When we begin to think of a conclusion as no longer being in question, our emotional attachment to it can reinforce confirmation bias and encourage us to protect the conclusion rather than test it.

Additionally, we should be cautious of any belief that requires us to reject vast networks of independently verified measurements simply because our unaided vision feels more trustworthy.

When someone thinks they have walked through a one-way door, the best way to test whether it truly is one-way is to see whether it can be opened again. We turn the handle and push or pull on the door to see if we can open it and walk back through. The way we do that with a belief is by examining the premises that support it—asking whether they are true and complete—and by checking whether the reasoning that led us to the conclusion is valid. If we discover a fallacy in our reasoning or in the premises, we have the amazing ability to humble ourselves and reverse beliefs we once thought were one-way.

## My Own Position

I have long been interested in conspiracy theories. I understand the instinct to distrust institutions, to look behind official narratives, and to suspect that things are not always as they seem. That instinct is not irrational. History contains real examples of deception and corruption.

I think distrust is a good starting point for investigation—but it can become an unreasonable filter that determines conclusions in advance. When distrust becomes the standard by which all evidence against a belief is judged, any evidence that contradicts our expectations can be dismissed—not because we have carefully examined it and found it wanting, but because it comes from a source we have already decided cannot be trusted. At that point, a belief or claim is no longer being tested. It is being protected.

Our approach to evaluating beliefs must be disciplined. We need to do more than test the claims of others—we need to test our own claims and beliefs by examining our supporting premises with the same care, and by being willing to say to ourselves, "Maybe my beliefs are wrong"—before deciding what the evidence shows.

It is easy to notice flaws in arguments that have conclusions we already disagree with.
It is much harder—and more important—to test and evaluate arguments with conclusions we currently believe, especially when those beliefs have become connected to our identity, certainty, reputation, or relationships with other people.

Confirmation bias is often strongest when we feel certain that our conclusion is already settled. At that point, contrary evidence can begin to feel less like information to evaluate and more like a threat to something we want to protect. For that reason, testing our own beliefs honestly is often more difficult—and more important—than criticizing the beliefs of others.

Because I already believed the spherical Earth model, I knew that if I wanted to practice the intellectual discipline I describe in this book, I needed to examine the strongest arguments against it. I did not want to assume my spherical Earth conclusion was correct simply because it was the conclusion I already held. I wanted to understand how a non-spherical model is said to explain the observations and measurements that I believed supported a spherical Earth.

At the same time, I recognized that if I was going to distrust the spherical model, which has survived centuries of measurement, cross-checking, correction, and refinement, then the burden was on me to determine whether a non-spherical model could explain at least as much, predict at least as well, and remain consistently verified across the same independent measurement systems.

In short, if I was going to change my mind, a non-spherical model had to explain at least as much—and do so at least as well—as the spherical model does.

For that reason, I chose a particular approach in this book. I did not want to argue about appearances or debate first principles, such as gravity, atmospheric pressure, or other foundational scientific topics, which often become points of endless disagreement.

Instead, I focused on the tool-obtained measurements and the mathematical calculations embedded in the technologies described in this book that enable them to function reliably in everyday life.

I was genuinely disappointed to discover that arguments against a spherical Earth often begin with something true: Earth appears flat. But they often stop there—at appearances, sometimes mistaking optical illusions for direct measurements and treating them as evidence that "simple math" proves the Earth cannot be a sphere.

Non-spherical Earth models are not coherent replacements and cannot account for many technologies without implicitly depending on the very spherical geometry they reject.

The systems examined in this book use spherical or spheroidal geometry, and their calculations must match measurable reality in order to function reliably. And they function extremely well.

Bridges and tunnels built toward the middle from both ends meet where calculated.
Aircraft arrive where their navigation systems calculate they will.
Radio signals propagate as engineers predict.
GPS receivers determine position with remarkable precision.

That all requires Earth's geometry to be that of a sphere.

## Judging the Models

Scientific models are not judged by any single observation. They are judged by how well they explain independent measurements across many fields.

The spherical Earth model explains those measurements consistently.

Non-spherical models do not.

Non-spherical thinking often isolates one visual appearance, one anomaly, or one distrust-based narrative from a larger body of consistent, independent measurements.

A single puzzling appearance cannot outweigh a network of independent measurements unless it explains that network better and more completely than the existing model.

## The Conclusion

It should now be more difficult for us to reject millions of consistent, tool-obtained measurements gathered over centuries and instead trust our visual imaginations and selective examples. The cost of holding a false belief is not just error—it is losing the ability to recognize the truth those measurements reveal. Rejecting so much established scientific knowledge because of one convincing argument is not careful analysis of evidence—it reflects an imbalance in how evidence is weighed.

Taken together, the independent measurements examined in this book do not merely suggest a spherical Earth—they describe one.

And those measurements are not abstract—they are the same systems you have just examined, tested, and compared.

If someone in the non-spherical community can clearly explain why independent measurement systems and the mathematics they use consistently indicate a spherical Earth when the Earth is actually flat, I am willing to evaluate whether that explanation accounts for the measurements as well as spherical geometry does.

Competing explanations have been proposed, but none have demonstrated the same level of consistent agreement across independent measurement systems.

If a non-spherical model were able to account for those measurements more accurately and consistently than the spherical model, I would revise my conclusion.

Until such an explanation is provided, however, the most reasonable conclusion is to trust what the measurements consistently indicate: Earth behaves as a sphere because the mathematics that describe those measurements are the mathematics of spherical geometry.

And until someone can show where those measurements—or the mathematics that describe them—fail, the most reasonable conclusion remains that
Earth is what those measurements consistently reveal:

a sphere.

## Closing

The argument presented in this book has concluded.

We now know:

Appearances describe what the world looks like from where we stand.
Measurements reveal what the world actually is.
And when they disagree, it is measurements—not appearances—that lead us to what is true.

Because appearances can include illusions—false imaginations that do not accurately represent reality—they cannot serve as the final authority over measurements.

In the first afterword that follows, I briefly address twenty common fallacies and false claims that often appear in discussions about the Earth's shape, followed by a short reflection on the importance of continuing to test our beliefs carefully.

# Afterword Part 1

## Reasoning Errors, False Claims, and Verifiable Reality

Even after strong evidence has been presented, disagreements often persist because some arguments feel persuasive while relying on subtle errors in reasoning—especially when appearances seem more convincing than measurements.

Before concluding this book, it is helpful to look directly at several of the most common reasoning patterns that appear in non-spherical arguments.

### Common Reasoning Errors in Non-spherical Claims

The reasoning errors listed below appear frequently in arguments against a spherical Earth. They are not signs of dishonesty or lack of intelligence. They are common, easy-to-make reasoning mistakes that arise when strong intuitions, incomplete models, and misplaced certainty interact. Recognizing these fallacies also helps us protect our minds from persuasive errors and keeps our emotions from deciding what we believe.

As you read through these, ask yourself whether you have ever used these forms of reasoning—or seen them used in arguments you found persuasive.

## 1. Misuse of Modus Tollens

***Description:*** Applying a valid logical form to a false or incomplete conditional premise.
***Why it feels convincing:*** The structure
- If A, then B
- Not B
- Therefore, not A

feels decisive and final.
***Why it fails logically:*** Modus tollens only produces reliable conclusions if its premises are complete and true. Long-distance visibility depends on factors including refraction, observer height, target height, and atmospheric conditions—not curvature alone
***Example:*** "If Earth were a sphere, distant cities would be hidden; we can see them; therefore, Earth isn't a sphere."

## 2. False Premise / Faulty Conditional

***Description:*** Using an incorrect or incomplete "If A, then B" statement.
***Why it feels convincing:*** It simplifies reality into an easy rule.
***Why it fails logically:*** If the conditional statement is false or incomplete, the conclusion does not follow from the premises alone.
***Example:*** "If Earth were a sphere, then the hump of its curvature would always hide distant objects from view."

## 3. Category Error

***Description:*** Applying rules from one system to another.
***Why it feels convincing:*** Familiar math feels transferable.
***Why it fails logically:*** Different systems require different models.
***Example:*** Treating Earth's curved surface as though it behaves exactly like a flat sheet of paper—such as in arguments involving spherical excess or navigation routes.

## 4. Equivocation

***Description:*** Shifting the meaning of a word during an argument.
***Why it feels convincing:*** The argument sounds consistent while quietly changing terms.
***Why it fails logically:*** Logical consistency requires stable definitions.
***Examples:*** Treating "level" as meaning flat rather than perpendicular to gravity; treating "horizontal" as meaning parallel everywhere rather than tangent to Earth's surface; treating "up" and "down" as absolute directions rather than directions defined by gravity; confusing "straight" with "geodesic," "flat map" with "flat Earth," or "water finds level" with "water forming a flat plane."

## 5. Argument from Personal Incredulity

***Description:*** Rejecting a claim because it feels unintuitive.
***Why it feels convincing:*** Intuition feels authoritative.
***Why it fails logically:*** Reality often contradicts intuition at large scales.
***Examples:*** "I don't feel motion, so Earth isn't moving." "There's nothing in space to push against, so rockets can't move or change direction in a vacuum. Therefore, spacecraft going to the Moon and back must be fake."

## 6. Argument from Ignorance

***Description:*** Claiming something is false because one does not understand it.
***Why it feels convincing:*** Unexplained things feel suspicious.
***Why it fails logically:*** Lack of understanding is not evidence of falsity.
***Example:*** "No one has explained gravity to me, so it doesn't exist."

## 7. Misunderstanding Perspective as Geometry

***Description:*** Confusing visual perspective effects with physical geometry *(not a formal fallacy name, but a useful description of a common reasoning error).*
***Why it feels convincing:*** Perspective is powerful and familiar.
***Why it fails logically:*** Perspective changes appearances, not physical relationships.
***Example:*** "Objects disappear bottom-up due to perspective, not curvature."

## 8. Cherry Picking (Selective Evidence)

***Description:*** Using only evidence that appears supportive.
***Why it feels convincing:*** It assures us our beliefs are correct.
***Why it fails logically:*** Emotional attachment to a belief can tempt us to ignore contrary evidence and bias our conclusions.
***Examples:*** Citing laser experiments over water while ignoring surveying, navigation, radio propagation, and GPS; highlighting unusually clear long-distance photographs while ignoring the many times distant objects disappear below the horizon; focusing on individual anomalies while ignoring thousands of routine tool-obtained measurements that agree with spherical geometry; showing selective Chicago skyline photos across Lake Michigan while ignoring measured elevations and atmospheric refraction.

## 9. Non Sequitur

***Description:*** Drawing a conclusion that does not logically follow from the premises.
***Why it feels convincing:*** The statements sound connected.
***Why it fails logically:*** The logical connection is missing.
***Example:*** "Curvature isn't obvious; therefore, Earth isn't curved." "Scientists have lied about COVID, so they are surely lying to us about Moon landings and the shape of the Earth too."

## 10. Overgeneralization

***Description:*** Generalizing from local or limited observations to global conclusions.
***Why it feels convincing:*** Repeated personal experience in controlled conditions feels decisive.
***Why it fails logically:*** Systems designed to operate locally do not reveal global geometry.
***Example:*** (aviation): "Pilots keep planes level throughout entire flights; therefore, they are not flying over a curved surface." (water behavior): "Water in lakes and canals appears flat, therefore the Earth must be flat." (gravity): "Objects fall straight down locally, so gravity should pull straight down everywhere." (trust): "Some scientists or institutions have been dishonest; therefore all scientists and scientific fields are untrustworthy."

## 11. Straw Man Fallacy

***Description:*** Attacking a simplified or distorted version of a theory rather than the real one.
***Why it feels convincing:*** The weakened version is easy to defeat.
***Why it fails logically:*** The real spherical model includes refraction, perspective, and instrument limits.
***Example:*** "The globe model says our line of sight should always be obstructed by a 66-foot hump of curvature at 10 miles."

## 12. Appeal to Conspiracy

***Description:*** A form of unfalsifiable reasoning that explains away conflicting evidence by invoking coordinated deception.
***Why it feels convincing:*** It protects our beliefs from falsification.
***Why it fails logically:*** It requires implausible global secrecy across many independent professions.
***Example:*** "Surveyors, pilots, astronauts, and engineers are all in on the great deception. They're all a bunch of stinkin' liars."

## 13. Contextomy (Quoting Out of Context)

**Description:** Selectively quoting or editing a statement in a way that removes its original meaning.
**Why it feels convincing:** A short clip or quote can appear clear and decisive when isolated.
**Why it fails logically:** Meaning depends on context. When the surrounding explanation is removed, the statement can be made to imply something the speaker never intended.
**Example:** Presenting a clipped statement from an astronaut or public figure to suggest that humans have not gone to the Moon, when the full statement actually refers to not going for a long time or to a different context entirely.

## 14. Moving the Goalposts

***Description:*** Changing standards of evidence when challenged.
***Why it feels convincing:*** It avoids admitting error.
***Why it fails logically:*** It makes claims unfalsifiable.
***Example:*** "Photos don't count; experiments don't count; NASA and Space Station videos are fake and don't count."

## 15. Appeal to Anonymous Authority (Hearsay)

**Description:** Relying on unnamed, secondhand, or unverifiable sources to support a claim.
**Why it feels convincing:** Statements attributed to insiders, experts, or professionals sound credible—especially when they suggest hidden knowledge.
**Why it fails logically:** Claims without identifiable, verifiable sources cannot be tested or examined. Scientific conclusions are not established through rumor or private testimony, but through publicly testable measurements.
**Examples:** "Pilots say they are not allowed to tell the truth about the shape of the Earth." "Admiral Byrd's diary describes flying

over a hidden land in Antarctica and seeing mammoths, so there must be unknown lands beyond Antarctica." "NASA scientists admit that the Earth is flat."

## 16. False Dichotomy

***Description:*** Presenting only two options when other possibilities exist.
***Why it feels convincing:*** Clear choices feel decisive.
***Why it fails logically:*** Rejecting one explanation does not automatically establish another.
***Examples:*** "If the globe model cannot perfectly explain every observation, then the Earth must be flat." Or, "If NASA has ever lied about anything, then the Earth cannot be a sphere." Or, "If space agencies use edited images, then space itself must be fake."

## 17. Model Rejection Without Replacement

***Description:*** Rejecting a theory without offering an explanatory alternative.
***Why it feels convincing:*** Disproof feels sufficient.
***Why it fails logically:*** Science requires the best explanation, not merely rejection.
***Example:*** "I am certain the globe model is wrong. It is not my job to provide a replacement."

## 18. Composition Fallacy

***Description:*** Assuming what's true locally must be true globally.
***Why it feels convincing:*** Local experience dominates perception.
***Why it fails logically:*** Whole systems behave differently from their parts.
***Example:*** "The ground looks flat here, so Earth is flat."

## 19. Instrumental Skepticism

***Description:*** Trusting instruments only when they agree with beliefs.
***Why it feels convincing:*** Selective trust feels cautious.
***Why it fails logically:*** It applies inconsistent standards of evidence.
***Example:*** Accepting amateur camera footage while rejecting GPS timing, satellite tracking, or live ISS video feeds: "ISS video of earth" live on YouTube.

## 20. Ad Hoc Rationalization

***Description:*** Inventing one-off fixes to save a belief.
***Why it feels convincing:*** It explains anomalies quickly.
***Why it fails logically:*** It fragments explanation and reduces predictive power.
***Examples:*** "Electrostatics replaces gravity," "Density and buoyancy replace gravity," or "Perspective causes bottom-up disappearance."

## False Claims and Verifiable Reality

Logical fallacies are not the only problem with many arguments against a spherical Earth. Another is that some commonly repeated claims about the world are simply not true.

Many of these claims are not based on direct evidence, but on references to documents or statements that are misrepresented or taken out of context. For example, it is often claimed that pilot training manuals or government and NASA documents "admit" that the Earth is flat. In reality, these documents do not state that the Earth is flat. They explain that, for the purpose of simplifying complex equations in flight dynamics, engineers sometimes assume a flat, non-rotating Earth as part of a set of mathematical approximations. These assumptions are explicitly introduced as simplifications—not as descriptions of physical reality. In actual navigation, aircraft systems use models that account for Earth's curvature and rotation to produce accurate routes and positions. Presenting those assumption statements as evidence that the Earth is flat is a misrepresentation of what the documents actually say. Claims built on statements taken out of context—or presented without their original purpose and meaning—cannot be treated as reliable evidence.

This kind of misrepresentation is often more persuasive than simple hearsay because it seems grounded in technical authority, but the original source does not support the claim being made.

For example, it is often claimed that Antarctica is an impenetrable wall of ice surrounding the edges of the world and that no one has ever circumnavigated the continent. In reality, sailors and research vessels have circumnavigated Antarctica many times, and long-distance sailing routes and races deliberately follow continuous paths around the continent—treating it as something to sail around, not a barrier enclosing the world.

Another claim frequently repeated is that no matter how high one travels above the Earth, the horizon always appears perfectly flat and curvature can never be seen. The author remembers observing curvature during a flight to Alaska in 1995—an observation he found both striking and awe-inspiring—and for that reason immediately recognized that claim as false. He is not the only one. Pilots, astronauts, high-altitude balloon operators, and many ordinary airline passengers report observing curvature at high altitude. Photographs and videos taken from weather balloons, research aircraft, and spacecraft consistently show the same curvature predicted by spherical geometry. That curvature is not produced by fish-eye lenses or by airplane windows. Please check this claim for yourself.

Misleading claims also appear in discussions about gravity. Some presentations suggest that gravity has never been measured, that it is merely a theory invented to explain motion, or that objects fall only because they are denser than the air or other surrounding fluids. In reality, gravitational acceleration has been measured with extraordinary precision for centuries, and modern instruments such as gravimeters and satellite data map variations in Earth's gravitational field across the entire planet.

Other commonly repeated claims include statements that GPS satellites are fictitious, that long-distance flight routes only make sense on flat maps, or that large bodies of water cannot follow a curved surface. Each of these statements contradicts well-documented observations from navigation, surveying, satellite tracking, and engineering practice.

When testing arguments about the shape of the Earth—or any scientific question—it is important not only to examine the logic of the arguments but also to verify whether the claims themselves are true. Do not accept claims about the way things are without checking them for yourself. Claims deserve belief only after surviving careful testing.

Beliefs built on false premises cannot reliably lead us to the truth. When a premise is shown by measurements to be false, the proper response is not to defend it by appealing to appearances but to revise any beliefs built upon it.

Before accepting a claim that appears to support a non-spherical Earth, it is important to confirm that it has not been taken out of context. A statement that sounds persuasive in isolation may say something different when read in full. Beliefs built on incomplete or misrepresented claims will lead to unreliable conclusions.

## The Unifying Philosophical Error

Perhaps the most serious problem with most non-spherical arguments is that they share a single, fatal weakness: a lack of explanatory and predictive coherence.

Non-spherical claims survive only by rejecting many technology systems at once—not because their measurements fail, but because what we see with our unaided vision can feel more trustworthy than tools, mathematics, and the accumulated work of millions of people across centuries.

But that is an imbalance in the weighing of evidence—often driven by emotional attachment to conclusions that feel more trustworthy than tools built to measure what our senses cannot reliably determine.

When asked how a non-spherical model might work, a common response is: "I don't know, and it's not my job to come up with a new model. But what I do know is that the Earth cannot be a sphere."

Rejecting the spherical Earth model without replacing it with another model that explains and predicts just as much does not solve the problem—it simply leaves the explanation incomplete. A claim that the Earth is "not a sphere" does no explanatory work by itself because it lacks the explanatory and predictive coherence required of a true model of reality. It does not tell us how surveying measurements remain consistent over large areas, how aircraft navigate long distances, how radio signals behave near the horizon, how satellites remain in orbit, or how GPS can locate us within meters.

Careful evaluation of our beliefs requires discipline: testing arguments, examining premises, resisting the emotional pull of conclusions we have grown comfortable defending, and guarding against confirmation bias by listening to contrary evidence.

We need this discipline because if a belief does not match reality, then something in what supports it—its premises, its reasoning, or its interpretation—must be mistaken, incomplete, or fallacious. And the only way to discover where that error lies is to continue testing those supports carefully and honestly.

Here is the hopeful part: it is not only easier, but far more freeing, to admit that we may have trusted an appearance too much or applied logic incorrectly—and then learn from that mistake. Letting go of a non-spherical Earth conclusion does not require letting go of God, meaning, or purpose. It requires only the humility to accept that the world is larger, stranger, and more beautiful than our immediate perspective suggests.

The spherical Earth is not a threat to wonder—it is an invitation to it. It is a world where oceans curve gently beneath our ships, where light bends through air, where satellites orbit in predictable harmony, and where solar and lunar eclipses can be timed centuries in advance. Trusting tool-obtained measurements and systems that agree mathematically does not diminish faith; it refines it. It replaces certainty built on illusions with awe that is grounded in discovery, and it allows us to move forward not in fear of being wrong, but in gratitude for having learned something true.

That is the standard we should apply whenever appearances and measurements seem to disagree.

## What Have We Learned?

Understanding these reasoning errors does not end the conversation—it clears the ground for thinking more carefully.

With the ground cleared, we can reflect on what this process has taught us about revising our beliefs in the pursuit of truth.

When our beliefs change, saying so openly is not a weakness—it is what gives our future claims credibility.

# Afterword Part 2

## Reflection: Truth, Humility, and Intellectual Discipline

At the beginning of this book, I told you that I consider myself a conspiracy theorist. That has not changed.

I still believe that powerful people conspire. I still believe governments, institutions, corporations, and intelligence agencies sometimes deceive the public.

But I also believe not every alleged conspiracy is real.

When we continue promoting a claim that has been shown to be false, others can recognize it—and when they do, it can influence how they evaluate everything else we say. If we are mistaken about something as basic as what can be objectively tested and measured, such as the shape of the Earth, our credibility on other issues is justifiably called into question.

If we want people to take us seriously when we talk about corruption, propaganda, censorship, or hidden power, we must be willing to test claims carefully—and avoid misleading others by presenting statements out of context or relying on incomplete or fallacious reasoning. Before sharing a claim, we should be sure we understand it fully and that it has been tested against evidence.

When I first heard the claim—from people I trust—that "simple math" disproves a spherical Earth, I took it seriously. It unsettled me. I assumed the people making that claim had already done the work to test it carefully. If the mathematics really ruled out a sphere, then my long-held belief would have to change.

However, as I began to listen to many non-spherical presentations, I realized it wasn't just one belief I would have to change—it was much more.

Accepting a non-spherical Earth is not revising a single belief—it requires rethinking a wide range of independently tested explanations: gravity, motion, light, air pressure, density, astronomy, the solar system, heliocentrism, the Sun and Moon, what a vacuum is, what the term "level" means, how we navigate, communication networks, and the technologies we rely on every day. In other words, it is a comprehensive shift in how we interpret the physical world—one that cannot be made responsibly without carefully testing every premise that leads us there.

When I realized that, I remembered a simple principle:
"Protect your mind.
Control your emotions."

That means protecting your mind from false or incomplete premises—and controlling your emotions so they do not reinforce confirmation bias before you have fully understood an argument.

Watching others debate this topic often led not to clarity, but to confusion and emotional escalation—because the goal too often becomes winning the exchange rather than discovering the truth. But winning—or convincing—is not the goal. Truth is.

Seeing how discussions quickly devolved into quarrels convinced me it would be better to write out the arguments and allow others to examine them carefully, at their own pace and in their own space—rather than in the middle of a heated "discussion."

Books allow us to slow down, set emotions aside, and examine each premise calmly. That is why this book exists.

And now, after working through the arguments and evidence presented here, I no longer believe any non-spherical Earth claims should be defended.
They should be laid to rest.

Saying that requires some clarification.

I do not mean appearances are meaningless.
I do not mean the non-spherical community is foolish.
And I do not mean the "simple math" argument is weak on first hearing.

Quite the opposite.

It is persuasive: it seems simple, logical, and familiar.

But arguments are not established by how convincing they feel—or by how many people they persuade.
They are established by whether their premises are true and complete, and whether their conclusions logically follow from those premises.
And those premises must be carefully tested—and shown to match reality.

What ultimately counts, then, is whether an argument is sound.
And if it is, then its conclusion is not a mere opinion—it is a logical consequence.

That is one of the central lessons of this book.

The "simple math" appearance-based argument fails because its premises are incomplete and treat an optical illusion as a direct measurement.
The measurements-based argument succeeds because its premises are supported across independent systems.

If we care about truth, then when an argument falsifies a belief we hold, the right response is not resistance—but reexamination. If the argument holds up under testing, we must be willing to change our minds.

The mind is most free when it is willing to be corrected by measurements and evidence, not when emotional attachment reinforces certainty and confirmation bias.

We are not machines. We want to be right. We can be embarrassed. We make mistakes. And when we are intellectually honest, we correct our mistaken beliefs. That is strength.

## Protect Your Mind. Control Your Emotions.

These ideas apply beyond this topic.
We need to test premises before we allow them to become beliefs—because a premise is just a premise until it has been tested, and most false beliefs begin as untested premises.

Practically, we can slow down when we hear claims like these:

- "Gas pressure always requires containment."
- "There is no outer space."
- "NASA buys helium—what do they need all that helium for but to keep all of their 'satelloons' up in the air?"
- "NASA has been lying about going to the big floating rock in the sky (Moon) for years."
- "No real images of satellites in space exist—only renderings."
- "You can see too far for the Earth to be curved."
- "Videos from the Artemis II mission were created using green screen technology—because NASA fakes everything."

Such claims can sound certain and shake our confidence, especially when they come from people who seem intelligent, informed, and trustworthy. But we should not let confidence substitute for evidence.

Instead, we should slow down and ask:

- Is that statement actually true and complete?
- What measurements support or refute it?
- Does it agree with the larger network of evidence?
- Does it require rejecting well-established scientific discoveries?
- Are all glitches in videos caused by on-location green screens—or could they result from lighting, compression, signal processing, or other technical factors?

A shocking statement is not evidence. A confident tone is not evidence. A viral clip is not evidence. Intelligent people can believe false things. Emotions can reinforce confirmation bias. So we need to control our emotions long enough to examine any claim carefully, objectively, and in full context.
And we can remember: it is better to be briefly embarrassed by correcting a mistake than to spend years defending a false belief.

## Humility Increases Credibility

If we have promoted non-spherical claims in the past, and we now see that they do not hold up, admitting that does not make us less credible. It makes us more credible.

It shows that we are capable of being corrected.
It shows that we care more about truth than preserving our intellectual image.
It reflects intellectual honesty—a willingness to recognize and admit when we have been persuaded by false claims or flawed arguments—and then choose to openly correct them.

People are far more likely to trust us when we say:
"I believed this. I tested it. I discovered it did not match reality. I changed my mind."

When we have that realization, publicly acknowledging that certain claims we shared were based on statements taken out of context or incomplete premises—and explaining why those ideas initially seemed persuasive but were later shown to be misleading—is not a weakness. It is a noble kind of honesty, and it requires real courage, especially if we have publicly defended those mistaken claims as more trustworthy than the spherical claims we now realize were true all along.

On the other hand, continuing to promote a claim after it has been tested and found to be unsupported risks undermining our credibility across other issues as well. That is one reason this flat-Earth or non-spherical claim should now be abandoned as false.

When we openly acknowledge that we were mistaken—and stop resisting the extensive body of evidence supporting a spherical Earth—our credibility is strengthened, not weakened.

We can also begin celebrating humanity's scientific discoveries and achievements, including Moon missions and much more, while continuing to test claims and humbly learn from experts across many scientific fields how things actually work, rather than assuming that millions of scientists, engineers, astronauts, and others are knowingly participating in a vast, coordinated deception.

## Being Persuaded Does Not Mean We Were Foolish

Changing our beliefs does not mean we were stupid or gullible. It means we were persuaded by an argument that felt powerful but rested on false or incomplete premises.

That can happen to any of us.
The answer is not shame.
It is humility and intellectual honesty.

We are easily persuaded this way not because we are foolish, but because the experience of seeing illusions while not knowing they are illusions is very convincing. We naturally believe what we see. The good news is that we can let false beliefs go and openly admit that our beliefs were wrong and then help others learn the truth. This ability is one of our greatest human superpowers.

This kind of honesty is rare and admirable.
And because it is rare, it is powerful.

## The Door Is Not One-Way

Some beliefs feel like one-way doors. Once we step through them, it can be emotionally difficult to go back.

But a belief only becomes a one-way door when we decide in advance that it cannot be wrong.
And a belief is not made true by becoming difficult to abandon.
It is made true by corresponding to reality.

One of the most freeing discoveries we can make is that a door we thought was one-way can, in fact, be opened.

We can go back.
We can revisit the premises.
We can test the reasoning.
We can admit that the argument no longer survives examination.

This is not defeat.
This is growth.

Again, this is one of our greatest human superpowers.

We can let false beliefs go—and when we do, we no longer need to protect them by explaining away contrary evidence.

## Wonder Is Not Lost When Error Is Corrected

In many ways, the truth is more awe-inspiring than an illusion.

Think about how remarkable it is that:
We humans can measure the world through surveying and discover its large-scale shape.

We humans can use the North Star or the Southern Cross to navigate across this enormous planet.
We humans can encode our voices into radio signals that carry messages across great distances.
We humans can use small devices to receive signals from satellites and determine our position anywhere on Earth.
We humans can launch equipment into space that orbits the planet and makes countless technologies possible.

All of that is remarkable.
This is not a dull world; it is an astonishing one.

Correcting a false belief about the shape of the Earth does not diminish wonder—it deepens it and frees us from needing to deny centuries of amazing scientific discoveries and achievements.

## After Testing the Argument

We are now in a position to judge what conclusion truly follows.

After evaluating the arguments presented in this book, I am where I began in one sense—and somewhere different in another.

I still tend to distrust official narratives.
I still believe conspiracies exist and that powerful people lie.

But after testing the "simple math" flat-Earth claim that first raised my doubts—and recognizing that its persuasive power comes from an optical illusion—and then testing many other non-spherical Earth claims against independent measurements, **I no longer believe any non-spherical Earth model can accurately describe reality.**

Whether you accept that conclusion is up to you.

What I hope you will do now is continue testing this—and other claims—and follow the evidence wherever it leads.

# Final Homework

As you continue testing claims for yourself, there are additional flat-Earth arguments you may want to examine.

To avoid confirmation bias, examine both the claims and the responses to them. Watch the *Level* documentary series first, allowing its claims to stand on their own. Then watch videos that critically analyze those claims using established scientific understanding.

As you do, ask: Which set of claims forms a more complete, internally consistent model? Which relies on complete premises, measured evidence, and systems that actually work? Which have been tested by independent measurements—and which have not? Which point more at appearances? Which at measurements?

The goal is to protect your mind and control your emotions. Compare claims fairly, test premises, ask the questions above, and follow the evidence where it leads.

**"Level (2021)"** Hibbeler Productions
youtube.com/watch?v=2hnrlyOvUcQ

**"The Next Level (2022)"** Hibbeler Productions
youtube.com/watch?v=3OfbwhU5PQk

**"Level with Me (2023)"** Hibbeler Productions
https://www.youtube.com/watch?v=dR47TRmLqcc

**"Debunked: SUPERCUT"** videos — SciManDan (YouTube)
*A flat-Earth debunker who uses science, reason, and logical thinking to evaluate such claims.*

**"Exposing Flat Earth Dave's Sales Pitch"** Dave McKeegan (YouTube), and any other videos McKeegan has created to challenge flat-Earth theories.

## Further Exploration: Fine-Tuning

In these discussions, the word "faith" is often misunderstood. It is sometimes defined as belief without evidence, but that definition is imprecise. *Faith* can be more accurately understood as *trust.*

Like trust in any other area of life, it is not blind. It is shaped by experience, evidence, and repeated confirmation. When we sit in a chair, we have faith that it will hold our weight—not because we believe without evidence, but because that expectation has been consistently supported by experience.

In the same way, some readers find that examining the evidence related to the fine-tuning of the universe leads them to place greater trust—what many would call faith—in the idea of an Intelligent Designer. Whether one approaches the evidence of fine-tuning from a scientific, philosophical, or theological perspective, exploring it further can deepen one's sense of wonder about the world we inhabit.

If the discussion in this book has led you to think more carefully about how we evaluate claims and follow evidence, you may also find it worthwhile to explore topics concerning the apparent fine-tuning of Earth and the universe.

For those interested in that line of inquiry, the Discovery Institute (**discovery.org/store/**) has published a number of books—particularly the Privileged Species series—that examine this fine-tuning from multiple lines of evidence.

I have read many of their books and found them thought-provoking and worth careful consideration.

The videos on the **"Discovery Science"** YouTube channel present evidence related to fine-tuning in an engaging and accessible way, inviting reflection on the possibility of an Intelligent Designer. Also, please watch their 2026 movie, "The Story of Everything."

## About the Author

Will York is a lifelong student of science, philosophy, and subjects related to the nature of reality and truth. He holds degrees in biology, education, and Christian apologetics, and in the 1990s taught high school biology and chemistry before later working in the film industry on music videos and television commercials.

At age thirty-four, Will was paralyzed in a mountain biking accident after misjudging the height of a drop-off, an experience that challenged his understanding of love and his beliefs about God and spirituality. Since then, he has spent more than twenty-five years in a wheelchair, much of that time coping with chronic pain by immersing himself in study, reflection, and countless audiobooks, lectures, podcasts, and educational material.

Will is especially interested in subjects that challenge assumptions and encourage people to define the words they use and to think carefully about truth, evidence, beliefs, and the nature of knowledge. His interests include science, philosophy, theology, health and nutrition, conspiracy theories, intelligent design, and the relationship between consciousness and reality. He is fascinated by controversial claims, not because he enjoys conflict, but because he genuinely wants to know what is true.

When his pain allows, he enjoys hand cycling, kayaking, boating, adventures with his wife (his "first reader," to whom he is very thankful), bird watching, creating art, and spending time in nature. He studied painting and sculpture at the Lyme Academy and deeply appreciates human ingenuity, creativity, and the amazing beauty and organized complexity of the natural world.

Will grew up in the Richmond, Virginia area and lives there today. He is hopeful that advances in treatments for spinal cord injuries will someday enable him to walk, run, and dance again.

Made in United States
Troutdale, OR
11/03/2026

56847384R00094